D1760892

KS3
Success

Science

Revision Guide

Dan Foulder

Age 11-14

CONTENTS

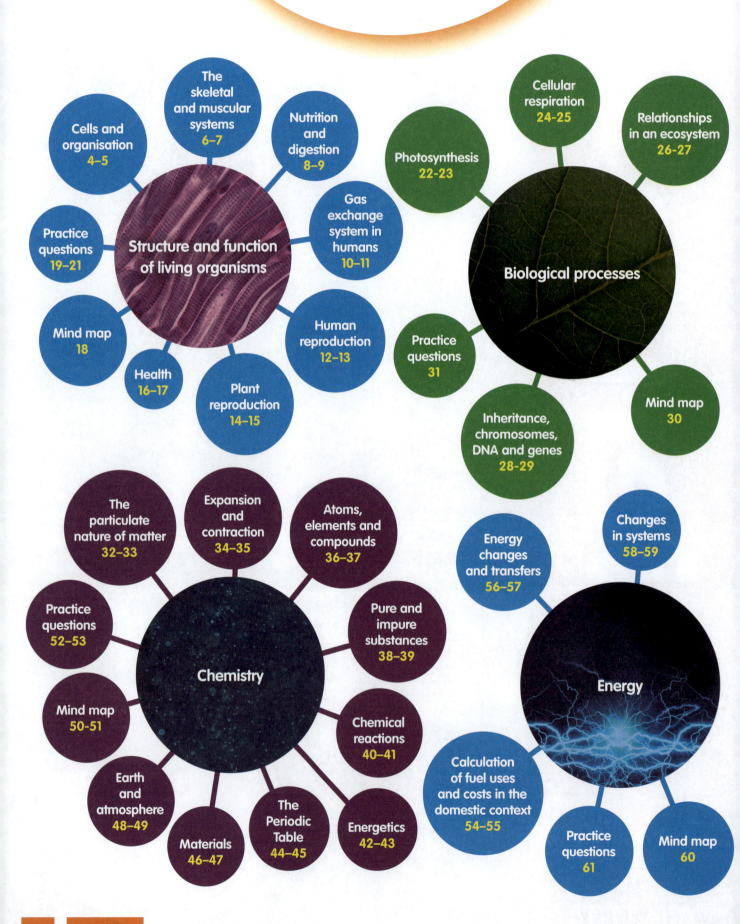

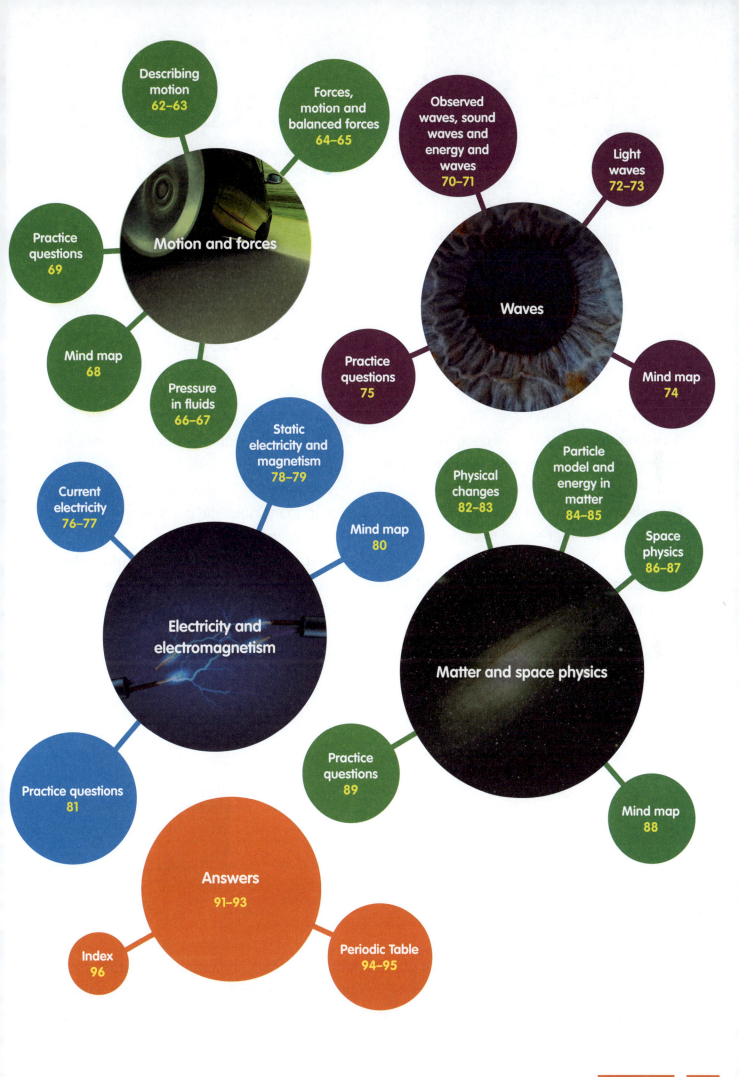

KEYWORDS

Multicellular (adjective) ➤ Organism made up of more than one cell.

Unicellular (adjective) ➤ Organism which is only made up of a single cell.

Diffusion (noun) ➤ Movement of substances from a high to low concentration.

Cells are the fundamental unit of living organisms. They are so small they can only be seen with a light microscope. Cells can be divided up into two main types: **animal cells** and **plant cells**.

Animal and plant cells have:

➤ A **cell membrane** that surrounds the cell and holds the contents in. It allows material to enter and exit the cell.

➤ **Cytoplasm** which is the jelly-like fluid of the cell. This is where chemical reactions occur in the cell.

➤ The **nucleus** which contains the cell's DNA (deoxyribonucleic acid) and controls the cell's functions.

➤ The **mitochondria** which carry out aerobic respiration.

Plant cells also have:

➤ A **cell wall** which prevents plant cells from bursting.

➤ **Vacuoles** which store cell sap. They also help to maintain the cell's shape and provide support for the plant.

➤ **Chloroplasts** which contain **chlorophyll** which is used in **photosynthesis**. Photosynthesis is the process by which plants produce glucose using energy from sunlight.

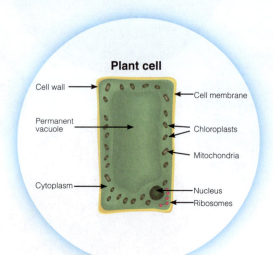

Materials are able to **diffuse** into and out of cells through the cell membrane. **Diffusion** is the movement of materials from an area of high concentration to an area of lower concentration. An example is oxygen diffusing from the cells of the lungs to red blood cells. Material can also diffuse within cells, e.g. glucose diffusing from the cytoplasm into the mitochondria during respiration.

A sponge is a simple multicellular organism

Make a model of an animal cell and a plant cell showing all the different parts of each. You can make your model out of paper or modelling clay, or you could be more creative and use jelly for the cytoplasm and a ping pong ball for the nucleus!

Unicellular organisms are specially adapted to survive. Some have **cilia** or **flagella** that allow them to move whilst other have eye spots which allow them to detect light.

Organisation

Cells are the most basic unit of living organisms.

Collections of similar cells working together to carry out the same function are known as **tissues**.

A group of tissues combine together to form an **organ**.

Organs form **organ systems**. The intestines and the stomach are part of the digestive system.

Organ systems interact to make up a complex multicellular **organism**.

Cilia

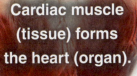

Cardiac muscle (tissue) forms the heart (organ).

1. What piece of equipment is needed to see cells?
2. What is the function of a chloroplast?
3. What do tissues group together to form?
4. Animal cells have cell walls. True or false?
5. What is the term for an organism that is only made up of one cell?

Humans have a skeleton comprised of 206 bones. The human skeleton is an internal skeleton. The skeleton has a variety of functions.

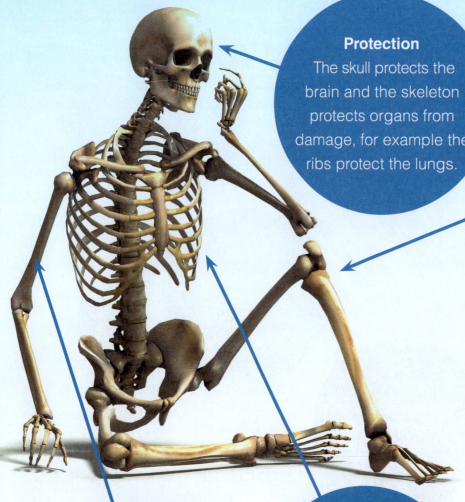

Protection
The skull protects the brain and the skeleton protects organs from damage, for example the ribs protect the lungs.

Movement
The skeleton provides an anchor point for muscles and allows the movement of limbs and other parts of the body.

Production of blood cells
Red and white blood cells are produced in the bone marrow.

Support
The skeleton supports the internal organs.

Biomechanics is the interaction of **muscles** and **bones**.

▼

Muscles are attached to bones by **tendons**.

▼

Muscles work by pulling on the tendons, which then pull on the bones.

▼

Muscles can only pull; they cannot push.

▼

When muscles **contract** they shorten and pull.

▼

When muscles **relax** they lengthen.

▼

One bone can pull on another bone, as bones are joined by **ligaments**.

KEYWORD

Antagonistic pair (noun) ➤
A pair of muscles which work in opposite directions to each other.

MODULE 2
THE SKELETAL AND MUSCULAR SYSTEMS

Muscles are often arranged into **antagonistic pairs**. The two muscles in an antagonistic pair work in opposite directions to each other. A good example of antagonistic muscles are the biceps and triceps, which raise and lower the forearm.

Experiments can be carried out to measure the force exerted by different muscles using a **force metre** (also known as a Newton metre).

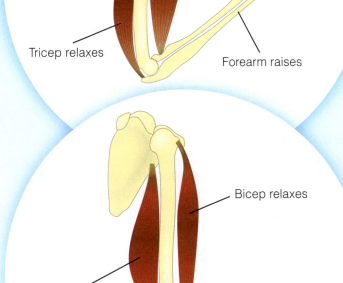

Bicep contracts

Tricep relaxes

Forearm raises

Bicep relaxes

Tricep contracts

Forearm lowers

Divide a large piece of paper into four equal-sized boxes. In each box draw a diagram to represent one of the functions of the skeleton, for example, for movement you could draw a picture of a person running to represent movement.

1. Muscles can only push not pull. True or false?
2. What is an antagonistic pair of muscles?
3. Give two functions of the human skeleton.
4. What piece of equipment can be used to measure the force exerted by muscles?
5. When a muscle relaxes does it shorten or lengthen?

MODULE 3
NUTRITION AND DIGESTION

Humans require a balanced diet in order to live a healthy life and maintain proper body function.

The components of a healthy diet:

Carbohydrates	➤ Used to release energy in respiration ➤ Excess carbohydrates which aren't used in respiration will be converted to fat for storage
Lipids (fats and oils)	➤ Used to store energy ➤ A diet too rich in lipids can lead to obesity
Proteins	➤ Used by humans for growth and repair of tissues
Vitamins and minerals	➤ Required in small amounts for the body to function
Dietary fibre 	➤ Important in providing roughage, which moves undigested food through the digestive system
Water	➤ Required by all cells ➤ Extremely important throughout the body, e.g. for transporting substances in the blood

Humans have a daily recommended intake of energy that varies depending on gender and level of activity.

➤ If a person does not get enough energy, they will find it difficult to concentrate.
➤ If a person regularly takes in too much energy, they will become overweight or suffer from obesity.

Imbalances in a human's diet can lead to a variety of health problems:

➤ **Obesity** – A diet that contains too much energy (usually too much carbohydrate and lipids), combined with a lack of exercise, can lead to obesity. The consequences of obesity include type 2 diabetes, high blood pressure and heart disease.
➤ **Starvation** – If a person is unable to get enough to eat they may begin to suffer from starvation. This may result in fatigue, muscle wastage and other long-term health problems.
➤ **Nutrient deficiency** – There are many conditions which are caused by deficiencies of specific **nutrients**, e.g. scurvy is caused by a **deficiency** in vitamin C and anaemia can be caused by iron deficiency.

Enzymes are **biological catalysts**, some of which speed up the breakdown of food in the digestive system.

Bacteria are very important in the human digestive system. They help in the digestion of various foods and also produce vitamins that humans require, such as vitamin K.

The digestive system

mouth – the food begins to be broken down in the mouth by the teeth, tongue and enzymes in the saliva

▼

oesophagus – the food travels down the oesophagus to the stomach

▼

liver – produces bile which neutralises stomach acid and helps the digestion of fats

▼

stomach – acid in the stomach kills bacteria that may have been on the food. There are enzymes in the stomach which speed up the chemical breakdown of the food

▼

small intestine – further chemical digestion by enzymes occurs in the small intestine. The products of digestion are absorbed in the small intestine

▼

large intestine – water is absorbed in the large intestine

▼

rectum / anus – undigested food passes out of the anus as faeces

Plants make glucose in their leaves by **photosynthesis**. Glucose is then converted into other organic molecules. Plants obtain mineral nutrients and water from the soil through their roots.

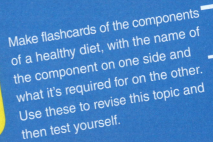

Make flashcards of the components of a healthy diet, with the name of the component on one side and what it's required for on the other. Use these to revise this topic and then test yourself.

1. What condition could be caused by eating foods with a high fat content?
2. Why are proteins important in the human diet?
3. How do plants obtain mineral nutrients?
4. What are the biological catalysts that speed up the digestion of food?
5. Where are the products of digestion absorbed?

MODULE 4
GAS EXCHANGE SYSTEM IN HUMANS

In humans the gas exchange system supplies **oxygen** to the blood and removes waste **carbon dioxide** from the blood.

🎧 4

1 Gases enter and leave a human through the nose and mouth.

2 Gases travel up and down the trachea, which contain rings of cartilage to prevent it collapsing.

3 The trachea branches into two bronchi (singular – bronchus).

4 The bronchi branch out to many bronchioles at the end of which are tiny sacs called alveoli.

5 The alveoli are very small sacs, with very thin walls and are surrounded by many capillaries. Oxygen diffuses from the alveoli into the blood in the capillaries. Carbon dioxide diffuses from the blood into the alveoli.

Nasal cavity
Nostril
Epiglottis
Larynx
Pharynx
Trachea
Primary bronchus
Right lung
Diaphragm
Left lung

Breathing in (inspiration)	Breathing out (expiration)
The external intercostal muscles contract and pull the ribs up and out.	The external intercostal muscles relax, the ribs move downwards and inwards.
The diaphragm contracts and flattens.	The diaphragm relaxes and raises upwards.
The lungs increase in volume.	The lungs decrease in volume.
The pressure in the lungs decreases.	The pressure in the lungs increases.
Air is drawn into the lungs.	Air is forced out of the lungs.

KEYWORDS

Inspiration (noun) ➤ Taking air into the body.
Expiration (noun) ➤ Forcing air out of the body.

Vital Capacity is the volume of air that can be taken into the lungs at one time. A person's vital capacity can be measured using a **spirometer**.

Breathing in and out

Air is drawn into the lungs when we breathe in (**inspiration**) and air is pushed out when we breathe out (**expiration**).

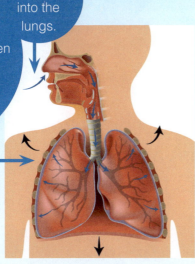

The air breathed in has a higher concentration of oxygen and a lower concentration of carbon dioxide.

Air is drawn into the lungs.

Pressure in the lungs is lowered by increasing the volume of the lungs.

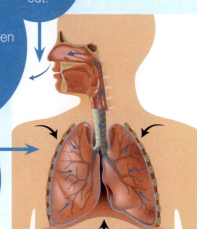

The air breathed out has a lower concentration of oxygen and a higher concentration of carbon dioxide.

Air is breathed out.

Pressure in the lungs is increased by decreasing the volume of the lungs.

Health factors

Different health factors can affect the functioning of the gas exchange system:

➤ **Exercise** leads to an increase in the volume of the lungs (vital capacity), allowing more oxygen to be taken into the blood.

➤ **Smoking** irritates the tissues of the gas exchange system, leads to build-ups of matter, coughs and potentially fatal conditions such as lung cancer and emphysema.

➤ **Asthma** causes the bronchioles to become constricted, reducing the volume of air that can get into the lungs. Asthma can be relieved using drugs taken with an inhaler. An asthma attack is an extreme and potentially life-threatening medical emergency.

Gas exchange in plants

Plants take in and release gases through small pores on the undersides of their leaves called stomata. During the day plants take in carbon dioxide and release oxygen. Plants require carbon dioxide for photosynthesis and oxygen for respiration.

The concentration of nitrogen remains the same during inspiration and expiration.

Write each of the stages of breathing in and out on separate cards (for example, ribs move downwards and inwards on one card, volume of lungs increases on another etc). Mix the cards up then arrange them into two piles, one for what occurs when breathing in (inspiration) and one for what occurs when breathing out (expiration).

1. How do gases enter the leaves of plants?
2. When breathing in, the ribs move in. True or false?
3. What is the name of the tube that carries gases from the mouth to the bronchi?
4. What gas diffuses into the blood from the alveoli?
5. What does the diaphragm do when a human breathes out?

MODULE 5
HUMAN REPRODUCTION

Humans reproduce by sexual reproduction. Like all mammals, **fertilisation** (fusing of the gametes) and development of the offspring occurs inside the female's body.

Gametes are the specialised cells used in reproduction. The male gamete is the sperm. The genetic material is in the head of the **sperm**. It uses its tail to swim to the female gamete.

The female gamete is the **egg**. The egg is much larger than the sperm and contains stored food for the developing embryo.

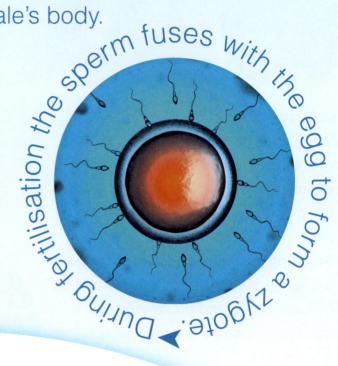

During fertilisation the sperm fuses with the egg to form a zygote. ◀

Female reproductive system

uterus – the sperm swim up through the uterus. After fertilisation has occurred the fertilised zygote will implant in the lining of the uterus (the endometrium)

oviduct – tube which allows the egg to move from the ovary to the uterus. Fertilisation occurs in the oviducts

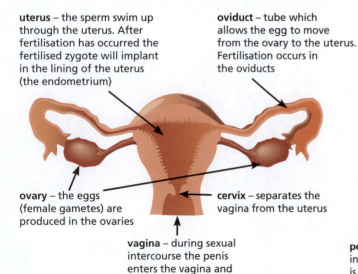

ovary – the eggs (female gametes) are produced in the ovaries

cervix – separates the vagina from the uterus

vagina – during sexual intercourse the penis enters the vagina and deposits the sperm

Male reproductive system

These diagrams are not to scale.

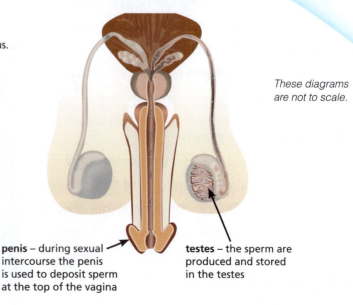

penis – during sexual intercourse the penis is used to deposit sperm at the top of the vagina

testes – the sperm are produced and stored in the testes

➤ When a human female starts **puberty** the **menstrual cycle** begins.

menstruation (3-7 days)

28 1 2 3 4 5 6 7 8 9 10 11 12 13 14 15 16 17 18 19 20 21 22 23 24 25 26 27

➤ Once a month, an egg is released from the ovaries and the lining of the uterus (the endometrium) thickens in preparation for implantation.

➤ If the egg is not fertilised, the endometrium breaks down and, along with the unfertilised egg, passes out of the female's body during menstruation.

ovulation

After nine months the mother will go into labour. Muscle contractions push the foetus out of the vagina and the baby is born.

Maternal lifestyle

The foetus requires substances such as oxygen and glucose, which are transferred from the mother's blood into the foetus' blood by the placenta. Harmful substances can also pass into the foetus, such as alcohol and heroin. If a mother takes these harmful substances, it can lead to growth problems and damage the developing foetus.

➤ If fertilisation does take place, the sperm will meet the egg in the oviduct.

➤ The sperm and egg fuse to form a **zygote**.

➤ The zygote divides to form an **embryo**. The embryo then moves down the oviduct to the uterus, where it implants in the endometrium. The embryo develops into a **foetus**. It is provided with nutrients from the mother's blood by the **placenta**.

Write the stages of human reproduction (fertilisation, implantation etc.) on separate small cards, mix the cards up and then put them back into the correct order.

5

1. What is the male gamete in humans?
2. Where does fertilisation in humans occur?
3. Where does the foetus develop?
4. How does the foetus obtain nutrition?
5. What is the name of the female gamete in humans?

MODULE 6
PLANT REPRODUCTION

Plants can reproduce sexually. Pollen contains the male gamete and must be transferred to the stigma of another flower. This is **pollination**. The pollen grain will then produce a pollen tube to allow the male gamete to reach the female gamete within the ovary of the plant.

The fertilised zygote forms a **seed**. The ovary forms the fruit around the seed. Fruits are used to help disperse the seeds.

Plant reproduction

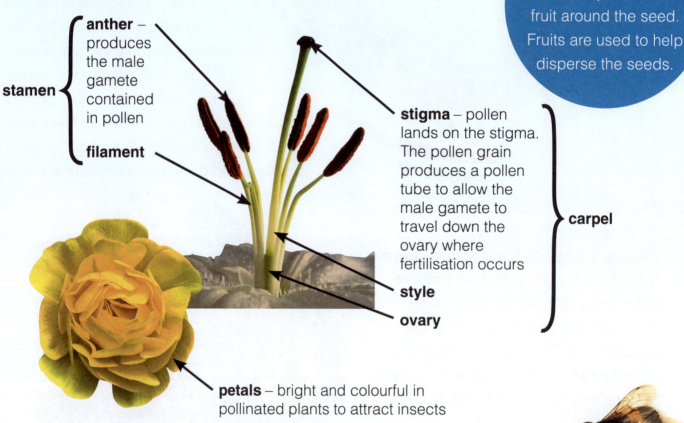

anther – produces the male gamete contained in pollen

stamen

filament

stigma – pollen lands on the stigma. The pollen grain produces a pollen tube to allow the male gamete to travel down the ovary where fertilisation occurs

carpel

style

ovary

petals – bright and colourful in pollinated plants to attract insects

There are two main methods plants use to transfer pollen:

➤ **Insect-pollinated flowers** are adapted to attract insects. They have bright petals and nectar. Pollen is sticky so it can stick to the insect's body.

➤ **Wind-pollinated flowers** are adapted to have their pollen dispersed by the wind. They don't have bright petals or nectar and their pollen is small and light.

Seeds can be dispersed in various ways.

Seeds can be adapted to be transported by the **wind**. Dandelion seeds are adapted for wind dispersal.

Seeds can either be transported internally or externally by **animals**. If an animal eats a fruit then the seeds may pass through the animal's digestive system and be deposited in faeces. Seeds may also have hooks on them to attach to the fur of animals and be carried away.

KEYWORD

Pollination (noun) ➤ The transfer of pollen from the anther to the stigma.

Seeds can be transported by **water**. This is particularly important in plants that live in water (aquatic plants) and those that live near to water

Take photos of different flowers you see in your garden/ neighbourhood. Divide the photos into those that seem adapted for wind pollination and those adapted for insect pollination.

1. What is pollination?
2. What happens to the fertilised zygote in flowering plants?
3. How do fruits help plant reproduction?
4. What type of flower would have small, light pollen?

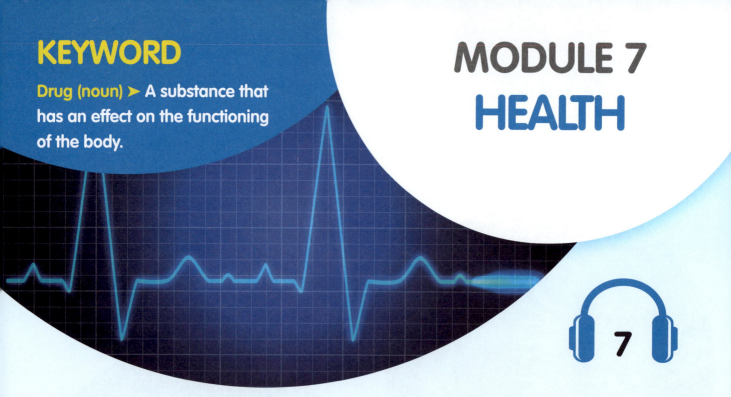

KEYWORD

Drug (noun) ➤ A substance that has an effect on the functioning of the body.

Recreational **drugs** have many negative health effects. Addiction to drugs not only affects the health of an individual, but is also a problem which affects society as a whole. Many drugs are addictive. This means users become dependent on the drug.

Illegal drugs

➤ **Cannabis** is derived from the cannabis plant and is smoked or eaten by users. Cannabis makes users feel very relaxed. It can also cause feelings of anxiety and paranoia. Cannabis may lead to long-term psychological problems.

➤ **Heroin** is an extremely addictive opiate. It suppresses pain. Some of the health problems associated with heroin are due to it being injected, often with dirty needles. Sharing dirty needles leads to the transmission of hepatitis and HIV. If a user takes too much heroin they can suffer from a fatal overdose.

➤ **Cocaine** makes users feel very alert and confident. Cocaine can be snorted through the nose or smoked in the form of crack cocaine. Cocaine is addictive. The health risks associated with cocaine include death from overdosing, heart attacks and other cardiac problems.

Snorting cocaine can lead to the cartilage in the nose being damaged.

➤ **Ecstasy** gives users more energy and makes them feel more alert. Ecstasy has been linked to heart, liver and kidney problems.

➤ **LSD** is hallucinogenic, which causes people to see and hear things that are not really there. Hallucinations can lead to dangerous accidents. The use of LSD can also lead to long-term mental health problems.

Make flashcards of the legal and illegal drugs on these pages. On one side of a card write the name of the drug. On the other side of the card write the negative health effects of the drug. Use your flash cards to revise and then test yourself.

Legal drugs

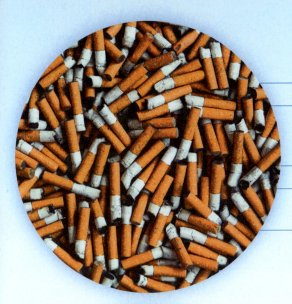

Smoking tobacco is legal in the UK.

Smoking tobacco can lead to lung conditions such as lung cancer and emphysema and can also cause heart disease.

Tobacco contains nicotine which is highly addictive.

Tobacco also contains tar which increases the risk of lung cancer.

Solvents include aerosols and glue.

Solvents can affect the heart, liver and kidneys.

The fumes from solvents are inhaled and can cause light-headedness.

Solvent abuse can lead to death from heart failure or suffocation.

Alcohol is legal in the UK and is generally considered socially acceptable. Too much alcohol will cause people to become drunk.

Long-term alcohol abuse may also to high blood pressure and heart disease.

Drunk people are at an increased risk of having an accident, particularly if a drunk person drives.

Long-term alcohol abuse leads to many health problems, including cirrhosis of the liver (a condition which damages the liver).

When people are drunk they are less in control of their actions and show poor decision-making.

1. Why does heroin addiction often lead to transmission of HIV and hepatitis?

2. Why could an LSD hallucination be dangerous?

3. How can cannabis be taken?

4. What are the dangers of solvent abuse?

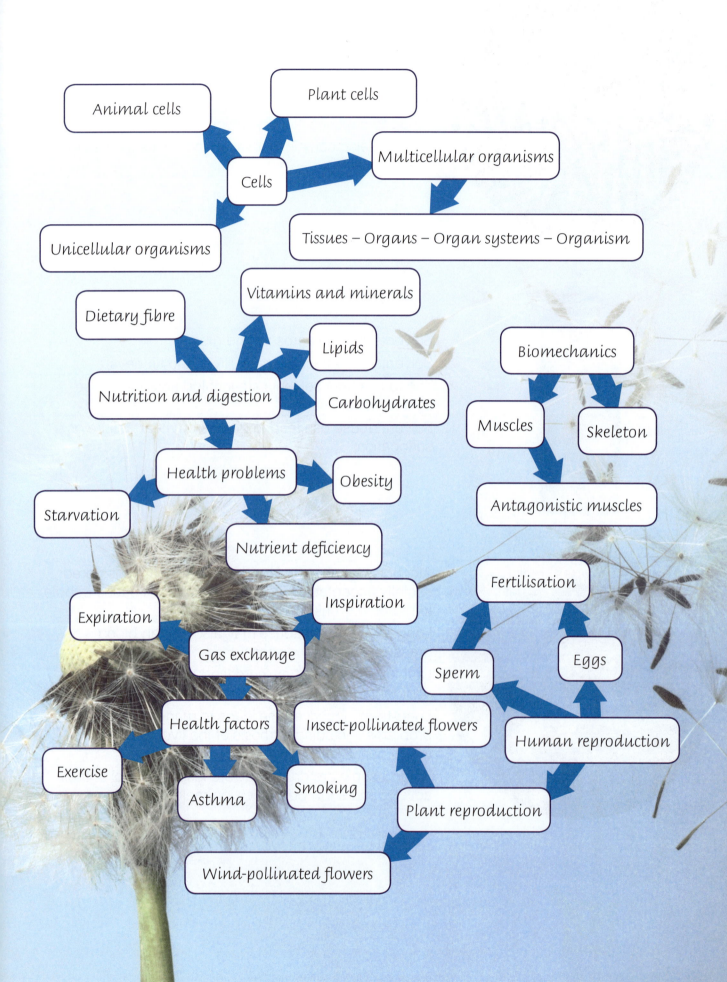

Animal cells

Plant cells

Cells

Multicellular organisms

Unicellular organisms

Tissues – Organs – Organ systems – Organism

Vitamins and minerals

Dietary fibre

Lipids

Biomechanics

Nutrition and digestion

Carbohydrates

Muscles

Skeleton

Health problems

Obesity

Antagonistic muscles

Starvation

Nutrient deficiency

Fertilisation

Inspiration

Expiration

Gas exchange

Sperm

Eggs

Health factors

Insect-pollinated flowers

Human reproduction

Exercise

Asthma

Smoking

Plant reproduction

Wind-pollinated flowers

STRUCTURE AND FUNCTION OF LIVING ORGANISMS
PRACTICE QUESTIONS

1. Steph and Lee were investigating the effects of exercise. They tested the force generated by the muscles in the arms of a variety of different people at a gym. They also recorded how many hours the person exercised per week. The results are shown below:

Hours of exercise per week	0–5	10–15	15–20	20–25	25–30
Average force generated by muscle (N)	80	100	140	190	250

a) Which of the following statements best explains this data? (1)

    A.   There is no relationship between the hours of exercise and force generated by muscles.

    B.   The people with the fewest hours of exercise were able to produce the most force.

    C.   Diet affects the force generated by the muscles.

    D.   The more hours exercised per week, the greater the force generated by the muscle.

b) The biceps and triceps were the muscles that moved the arm in this experiment. Complete the sentences below. (3)

The biceps and triceps are _____ muscles as they work in opposite directions. When the bicep _____ it shortens. At the same time, the tricep _____ .

c) The muscles are attached to the skeleton. One of the main functions of the skeleton is to allow a human to move. Give two other functions of the skeleton. (2)

2. Dave and Aqsa were examining two different flowers. Below are their notes:

A. Has large, brightly coloured petals and produces a strong scent.

B. Has small, dull-coloured petals and contains large amounts of light pollen.

a) What can be concluded about pollination in flowers A and B? (2)

b) On the image below label where pollen is produced and where pollen lands to germinate. (2)

c) The seeds of flower A have many small spines. What type of dispersal is this seed adapted for? (1)

3. Hannah and Yasmin are investigating cells. The cell they are studying contains a nucleus and mitochondria. Hannah thinks this proves the cell they are studying is a plant cell.

   a) Do you agree with her? Explain your answer. (2)

   b) Name two features which are only found in plant cells. (2)

   c) Yasmin believes the samples they are studying came from a multicellular organism. What is a multicellular organism? (1)

4. Alice and Lin are studying diet. They have made a list of the things needed for a balanced diet:

   ➤ Lipids

   ➤ Vitamins and minerals

   ➤ Water

   a) Complete their list. (3)

   b) Lipids (fats and oils) are an important part of a balanced diet, but a diet that is too high in fat can lead to health problems. What condition can be caused by eating foods high in fat? (1)

   c) Lipids are broken down chemically by enzymes. What are enzymes? (1)

5. a) Which of the following statements on gas exchange in humans is correct? (1)

    A.   The gas breathed out has a higher concentration of carbon dioxide than the gas breathed in.

    B.   The gas breathed out has a higher concentration of oxygen than the gas breathed in.

    C.   The gas breathed out has the same concentration of carbon dioxide as the gas breathed in.

b) Complete the following sentences on gas exchange in humans. (4)

Inspiration is breathing _____. During inspiration the volume of the lungs _____. This causes the pressure in the lungs to _____. In order to breathe out the pressure in the lungs must _____

c) Below are factors which can affect how well the gas exchange system in humans works. Circle the factors which will have a negative effect on the gas exchange system. (2)

Running four times a week.

Smoking five cigarettes a day.

Swimming three times a week.

Having asthma.

Plants and algae produce glucose and other organic compounds by **photosynthesis**. **Carbon dioxide** and **water** are the reactants and the source of energy is **light**. **Oxygen** is also produced.

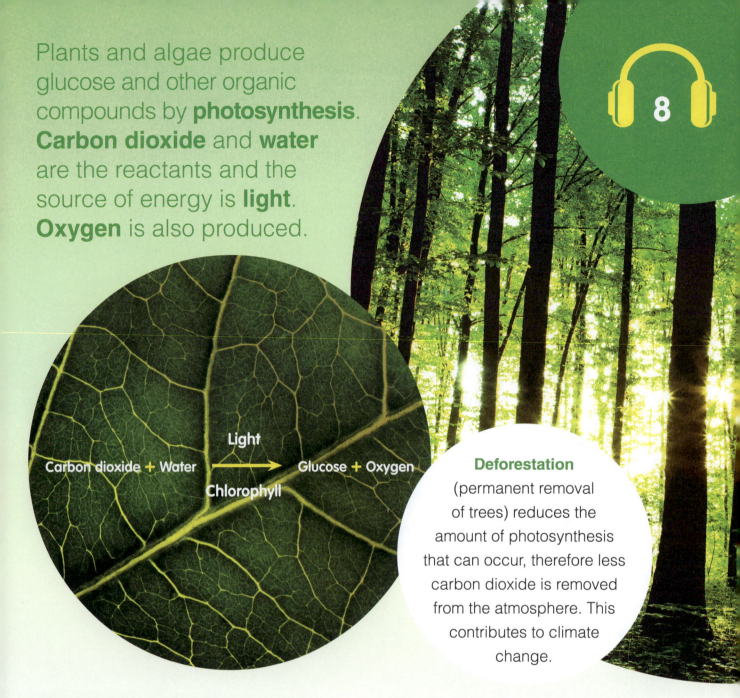

Light

Carbon dioxide + Water → Glucose + Oxygen

Chlorophyll

Deforestation
(permanent removal of trees) reduces the amount of photosynthesis that can occur, therefore less carbon dioxide is removed from the atmosphere. This contributes to climate change.

Photosynthesis occurs in the chloroplasts within plant cells. The chemical **chlorophyll** (a pigment) is used in photosynthesis.

Nearly all life on Earth relies on the organic compounds that are produced by plants and algae in the process of photosynthesis. **Herbivores** (plant-eating animals) eat these products directly and utilise the energy they contain. **Carnivores** (meat-eating animals) are then able to feed on the herbivores and utilise the energy they contain.

Over billions of years plants and algae have altered the Earth's atmosphere.

They have increased the levels of oxygen in the atmosphere and decreased the levels of carbon dioxide.

Today, plants help maintain the concentrations of carbon dioxide and oxygen in the atmosphere.

Create a comic strip summarising the importance of photosynthesis to all living organisms, both in providing organic compounds and taking in carbon dioxide from the atmosphere.

MODULE 8
PHOTOSYNTHESIS

(noun) ➤ The process by which carbon dioxide is fixed into glucose using light as a source of energy. Oxygen is also produced.

KEYWORDS
Chlorophyll (noun) ➤ Green chemical used in photosynthesis.
Leaf (noun) ➤ Main photosynthetic organ of a plant.

The main photosynthetic organs in plants are **leaves**. They have a large surface area to absorb as much light as possible. Many plants have leaves which are also able to orientate (turn) themselves to the sun to ensure they absorb the most light possible.

Leaf Anatomy

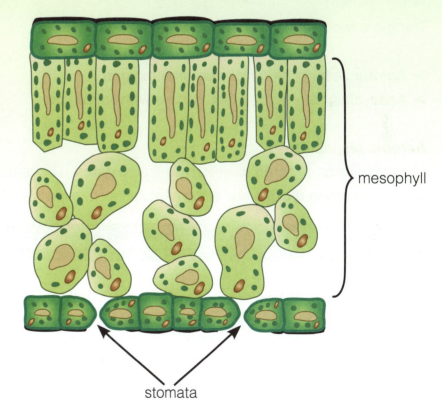

mesophyll

stomata

➤ Small pores called **stomata** on the bottom of leaves allow carbon dioxide to enter the leaf for photosynthesis and for the oxygen produced to be released from the leaf.

➤ The majority of photosynthesis occurs in the **mesophyll** cells.

➤ **Xylem** provide the water for photosynthesis and **phloem** are used to transport the products of photosynthesis away from the leaf.

1. What gas is a reactant in photosynthesis?
2. What gas is a product of photosynthesis?
3. Where in the cell does photosynthesis occur?
4. What green chemical is required for photosynthesis?

MODULE 9
CELLULAR RESPIRATION

All living organisms require **energy** to survive. Energy is released by breaking down **glucose** in **respiration**. The energy released in respiration allows all other chemical process necessary for life to occur.

There are two types of respiration:

➤ **Aerobic** respiration, which **requires oxygen**.
➤ **Anaerobic** respiration, which occurs **without oxygen**.

Aerobic respiration

Most respiration is **aerobic**, which requires **oxygen**. The reaction releases **energy** and produces **carbon dioxide** and **water**. Aerobic respiration is shown in this word equation:

Glucose + Oxygen ⟶ Carbon dioxide + Water + Energy

Anaerobic respiration

If there's insufficient oxygen available, energy can still be released by **anaerobic respiration**. Anaerobic respiration is much less efficient than aerobic respiration, so produces less energy. Most cells can only survive for a short time on the limited amount of energy produced in anaerobic respiration.

In **animal cells** anaerobic respiration produces **lactic acid**. During intense exercise, muscles don't receive enough oxygen, so muscle cells have to respire anaerobically. The burning sensation felt in muscles during intense exercise is due to a build-up of lactic acid.

Glucose ⟶ Lactic acid + Energy

Respiration (noun) ➤ The process by which living organisms release energy from organic molecules such as glucose.
Aerobic (adjective) ➤ Occurs when oxygen is present.
Anaerobic (adjective) ➤ Occurs when no oxygen is present.

In **microorganisms** (e.g. **yeast**) anaerobic respiration produces **alcohol (ethanol)** and **carbon dioxide**. Anaerobic respiration in yeast is also known as **alcoholic fermentation**.

Aerobic respiration vs Anaerobic respiration

	Aerobic respiration	Anaerobic respiration
Reactants	Glucose Oxygen	Glucose
Products	Water Carbon dioxide	Lactic acid in animal cells Alcohol and carbon dioxide in microorganisms
Energy	Releases a lot of energy	Releases less energy than aerobic respiration

Glucose ⟶ Alcohol + Carbon dioxide + Energy

Make small cards of all the reactants and products in both aerobic and anaerobic respiration. Mix them up and then arrange them to give the correct equations for both aerobic respiration and anaerobic respiration.

1. What gas is required for aerobic respiration?
2. What gas is released in aerobic respiration?
3. In what type of organism is alcohol produced in anaerobic respiration?
4. Which type of respiration is most efficient?

MODULE 10
RELATIONSHIPS IN AN ECOSYSTEM

The feeding relationships between organisms can be shown by food chains.

In a food chain, arrows show the flow of energy:

10

All food chains start with a **primary producer**, a green plant, which produces organic compounds by **photosynthesis**.

The organism which feeds on the primary producer is a **primary consumer**, also called **herbivores** as they feed on plants.

The food chains in an ecosystem can be linked together to form more complex **food webs**.

Organisms which feed on primary consumers are **secondary consumers (carnivores)**. Organisms which feed on secondary consumers are **tertiary consumers**.

At each stage of a food chain energy is lost as **heat** and **waste**.

Insects are very important in plant reproduction. They transfer pollen from one flower to another to allow plants to sexually reproduce.

Many crops eaten by humans are pollinated by insects. It is important that we preserve populations of insects to ensure our own **food security**.

Organisms can affect their ecosystem and are affected by it. The effect of **toxic materials** in ecosystems is a big concern for conservation.

Toxic materials can poison organisms. The effects can be short-term, instantly effecting organisms. Toxic materials that are **persistent** accumulate in the food chain and can affect consumers further up the chain for a long period of time.

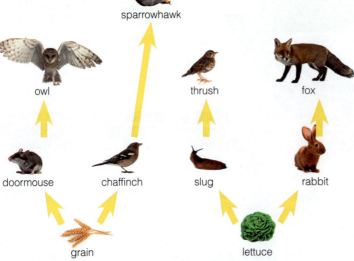

sparrowhawk

owl

thrush

fox

doormouse

chaffinch

slug

rabbit

grain

lettuce

How toxic materials can accumulate in ecosystems include:

Heavy metals, e.g. mercury and lead contaminate water.

> Simple marine organisms (such as plankton) can take in heavy metals if the water they live in is contaminated with them.

> Fish feed on plankton and the heavy metals are then transferred to the fish.

> The heavy metals build up in the tissue in the fish.

> The fish will suffer from heavy metal poisoning.

> If a human eats the fish they will also suffer from heavy metal poisoning.

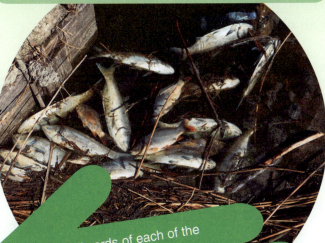

Primary producer (noun) ➤ An organism which produces its own food. Primary producers are the basis of food chains.

Consumer (noun) ➤ An organism which feeds on another organism.

Pesticides are used to kill pests that damage crops and other human interests, e.g. DDT is a pesticide.

> DDT is sprayed onto crops.

> Insects feed on the crops and take in the DDT.

> Small birds feed on the insects and the DDT is transferred to them.

> A larger bird, e.g. a hawk, eats the small birds.

> A large amount of DDT builds up in the hawk's body.

> The DDT causes the shells of the bird's eggs not to form properly, meaning they break before they can hatch. This reduces the bird's population.

DDT is now a banned substance.

Make small cards of each of the organisms in the food web on this page. You can either draw a picture of the organism or just write its name on the card. Mix the cards up and without looking in this book, try to arrange the cards to show the correct food web.

1. Give an example of a primary producer.
2. In a food chain a zebra eats some grass and is then eaten by a lion. What term could be given to the zebra?
3. In the food chain in question 2, what term could be used to describe the lion?
4. Give two examples of toxic materials which can accumulate in food chains.

MODULE 11
INHERITANCE, CHROMOSOMES, DNA AND GENES

11

Genetic information is transmitted from one generation to the next. This is **heredity**.

Deoxyribonucleic acid (**DNA**) is the molecule which carries the genetic information. DNA is divided up into **genes**. Each gene is the genetic code for one protein. DNA is contained in **chromosomes**.

Organisms that can breed together to produce fertile offspring are members of the same **species**. Organisms which are not the same species cannot breed together to produce fertile offspring. This is because their DNA is too different. African lions (*Panthera leo*) are a different species from tigers (*Panthera tigris*), and therefore they cannot breed together to produce fertile offspring.

Within a species there is **variation**. Variation can either be continuous or discontinuous.

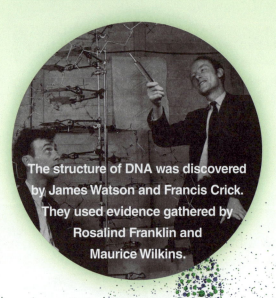

The structure of DNA was discovered by James Watson and Francis Crick. They used evidence gathered by Rosalind Franklin and Maurice Wilkins.

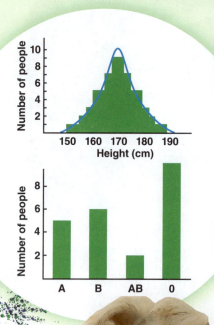

Continuous variation can take any value between a maximum and a minimum, e.g. height.

Characteristics which can be put into categories, e.g. blood group, show **discontinuous variation**.

Humans have 46 chromosomes arranged in 23 pairs.

If an environment changes, all the members of a species may not be well adapted to it. If a species is unable to successfully compete and reproduce, it will become **extinct**. A species is extinct when it has no living members, e.g. dinosaurs. It is thought the dinosaurs became extinct due to changes in their environment caused by a meteor strike on Earth and the climate change this caused.

Human activity has accelerated the rate of species extinction and many people are worried about the organisms that are being lost forever. Extinction leads to a reduction in **biodiversity** (the number of different species in an area). Loss of biodiversity could have many negative impacts, including affecting human food supplies if crops aren't pollinated by insects which have become extinct.

Conservation works to prevent extinction and maintain biodiversity. **Gene banks** can also keep samples of animal embryos and plant seeds that could be used to reintroduce organisms if they do become extinct.

There is competition between members of the same species and between different species for survival.

Due to variation, some members of a species will be better adapted to survive than others.

The organisms that are best adapted will survive and pass on their well adapted characteristics to their offspring.

The offspring are then also more likely to be better adapted and survive. This is natural selection.

Using the flow chart on this page, write the stages of natural selection on small cards, mix them up and then place them in the right order.

1. What is the term for organisms being well adapted, reproducing and passing on their beneficial characteristics to their offspring?
2. If two organisms can breed together to produce fertile offspring they are the same ...?
3. Weight is an example of which type of variation?
4. What is heredity?
5. Which four scientists were involved in the discovery of the structure of DNA?

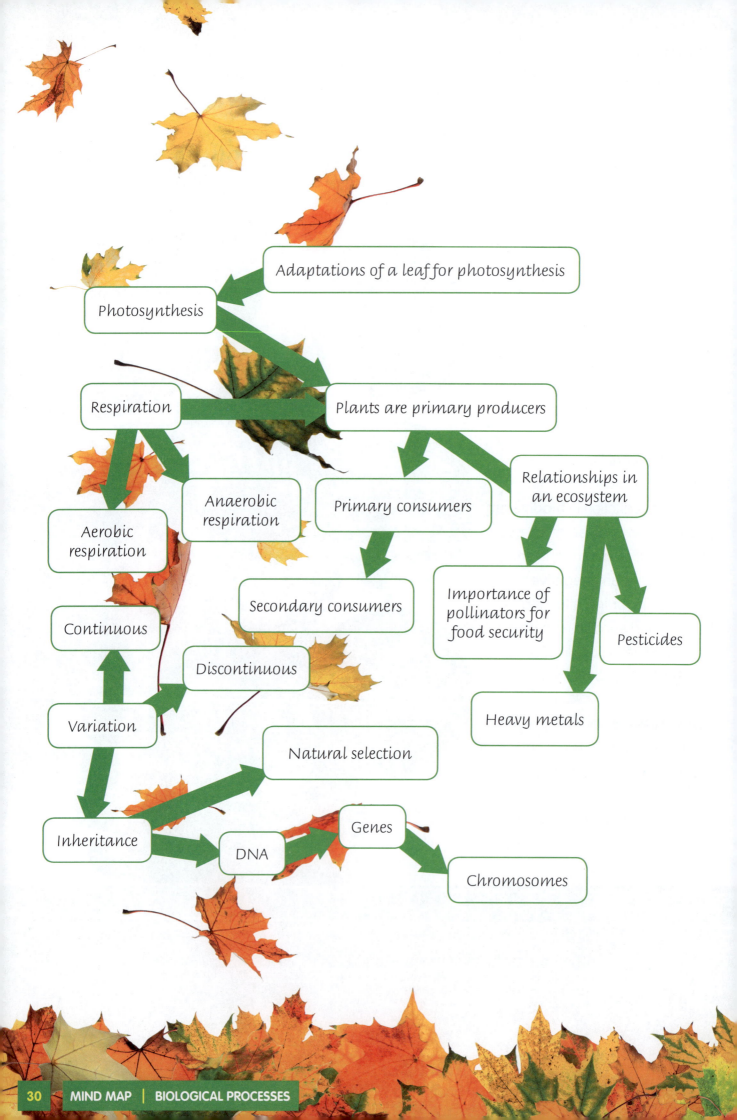

Adaptations of a leaf for photosynthesis

Photosynthesis

Respiration

Plants are primary producers

Anaerobic respiration

Primary consumers

Relationships in an ecosystem

Aerobic respiration

Secondary consumers

Importance of pollinators for food security

Pesticides

Continuous

Discontinuous

Heavy metals

Variation

Natural selection

Inheritance

DNA

Genes

Chromosomes

BIOLOGICAL PROCESSES
PRACTICE QUESTIONS

1. a) In the food web opposite identify:

    A.   a primary producer

    B.   a primary consumer

    C.   a secondary consumer (3)

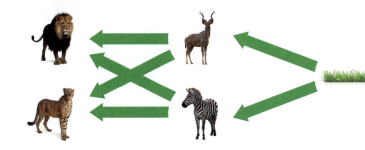

   b) Jake is learning about bees. He believes that it's a good thing that the bee population is decreasing, as it means less people will be stung. Why should Jake be concerned about the decline in the bee population? (2)

   c) Scientists are worried that some species of bee may eventually become extinct. What is meant by the term extinction? (1)

2. The graph below shows the beak length of a species of bird.

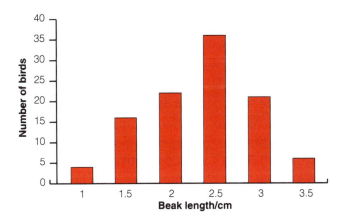

   a) What type of variation is beak length? (1)

   b) Complete the sentences below to show how the birds may have evolved to have a long beak.

   There is competition between birds of the same species. Birds with longer beaks have a better chance of surviving and passing on the genes for big beaks to their offspring. This is an example of _____ _____ (1)

   c) What are the 'big beak' genes made up of? (1)

   d) Scientists are very worried that this species of bird is on the edge of extinction.

      How could gene banks help in the conservation of this bird? (1)

MODULE 12
THE PARTICULATE NATURE OF MATTER

All **matter** is made up of **particles**. The behaviour of these particles determines the 'state' of the matter. All particles move – when a particle is heated, it gains **kinetic energy** and moves faster.

There are three **states of matter**:

Solids	Liquids	Gases
The particles in a solid are very close together.	The particles in a liquid are close together.	The particles in a gas are not close together.
Strong forces of attraction hold the particles together.	The forces of attraction between the particles in a liquid are not as strong as the forces of attraction between particles in a solid.	The forces of attraction between the particles are weak.
The particles in a solid vibrate, but don't move away from each other. Their positions are fixed.	The particles in a liquid can move relative to each other. This allows liquids to flow.	The particles in a gas move rapidly in all directions. This means that gases will expand to fill a space.
Solids have a fixed volume and shape and are difficult to compress.	As liquids flow, they do not have a fixed shape (they take the shape of the container they are in). They have a fixed volume and are difficult to compress.	Gases do not have a fixed shape or volume and are easily compressible. When a gas is compressed its pressure increases.

Pressure

Gas particles move quickly in all directions. If they are in a container, they will hit the sides of the container they are in and this causes pressure.

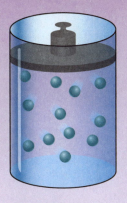 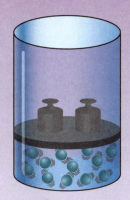

Lower pressure   **Higher pressure**

Pressure can be increased by:
➤ Heating the gas.
➤ Compressing the gas
(decreasing the size of the container).

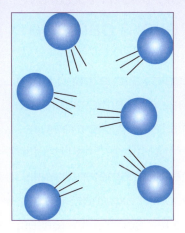 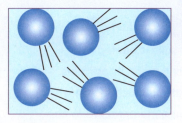

If the gas particles are heated they will move around faster. This means they will strike the side of the container with more force and more often, therefore increasing the pressure.

If the size of the container is decreased the gas particles will hit the sides of the container more often than they would in a larger container. This will cause the pressure to increase further.

KEYWORDS

Kinetic energy (noun) ➤ the energy a particle has when moving.
States of matter (noun) ➤ the different forms matter can take: solid, liquid or gas.

Write down each of the properties of a solid, liquid and gas on separate cards, for example, 'fixed volume'. On a large piece of paper, draw three boxes. Label one box 'Solids', one box 'Gases' and one box 'Liquids'. Place each of the cards in the box that has that property. Some cards may fit in two boxes!

1. Which of the three states of matter has a fixed shape and volume?

2. Which of the three states of matter does not have a fixed shape or volume?

3. Which of the three states of matter have the strongest attractive forces between the particles?

4. If a gas is heated what will happen to the pressure?

5. If the volume of the container a gas is in increases, what will happen to the pressure of the gas?

KEYWORDS

Change of state ➤ When a substance changes from one state of matter to another, for example, from a solid to a liquid by melting.

When a solid is heated it will expand and when it is cooled it will contract. This has lots of practical consequence, for example, sections of railway tracks have to have gaps between them to allow the metal of the track to expand on hot days to prevent damage.

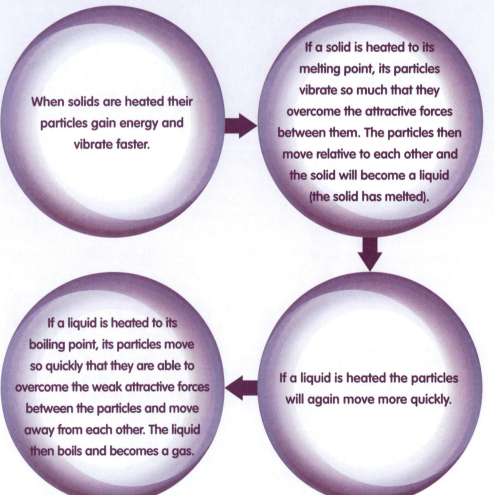

When solids are heated their particles gain energy and vibrate faster.

If a solid is heated to its melting point, its particles vibrate so much that they overcome the attractive forces between them. The particles then move relative to each other and the solid will become a liquid (the solid has melted).

If a liquid is heated to its boiling point, its particles move so quickly that they are able to overcome the weak attractive forces between the particles and move away from each other. The liquid then boils and becomes a gas.

If a liquid is heated the particles will again move more quickly.

Liquids also expand when heated. Thermometers use this principle to measure temperature. The liquid in the thermometer (usually alcohol) expands when the temperature increases, rising up the thermometer. When the liquid cools, it contracts and moves down the thermometer.

MODULE 13
EXPANSION AND CONTRACTION

When matter cools, the particles lose energy and move slower.

When a gas is cooled below its boiling point, the particles no longer have the energy to overcome the attractive forces. The gas becomes a liquid (it condenses).

When a liquid is cooled below its melting point, the particles no longer have enough energy to move around relative to each other. The liquid becomes a solid (it freezes).

Write out each of the stages of the flow charts from this module on small cards. Mix all the cards up and arrange them into the correct sequences to show:
- The heating of a solid to form a gas;
- The cooling of a gas to form a solid.

1. When a solid melts to form a liquid, do the particles gain energy or lose energy?
2. How can a gas be converted into a liquid?
3. What happens to particles of matter when they are cooled?
4. When will a solid expand?
5. When will a liquid contract?

MODULE 14
ATOMS, ELEMENTS AND COMPOUNDS

An **atom** is the simplest unit of matter. Atoms are made up of three subatomic particles:

| Protons | Neutrons | Electrons |

Protons and **neutrons** are found in the nucleus at the centre of an atom. Electrons orbit the nucleus. The particles in an atom have different masses and charges.

Particle	Relative mass	Charge
Proton	1	Positive
Neutron	1	Neutral
Electron	1/1800	Negative

Elements are substances that are made up of only **one type of atom**. The elements are shown in the **periodic table**.

Elements

Elements are represented by symbols in the periodic table. Some of them are easy to identify (for example, H for hydrogen, N for nitrogen), but some are harder (Au for gold). The first letter of an element symbol is a capital and the second letter (if any), is lower case.

A **compound** is made up of atoms of more than one type of element. Water is an example of a compound, because it is made up of two atoms of hydrogen and one oxygen atom. It therefore has the formula H_2O (H for hydrogen and O for Oxygen).

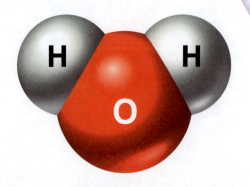

Make a model of an atom showing the protons and neutrons in the nucleus and the electrons orbiting the nucleus. You could use bits of paper or modelling clay.

H   H

O

100g Ice

100g Liquid water

KEYWORDS

Atom (noun) ➤ Basic unit of matter consisting of protons, neutrons and electrons in the nucleus.
Element (noun) ➤ Substance made up of only one type of atom.
Compound (noun) ➤ Substance made up of more than one element.

Mass is always conserved in **changes of state**. When a solid melts to become a liquid, the mass of the liquid will be the same as mass of the original solid.

250g Products

250g Reactants

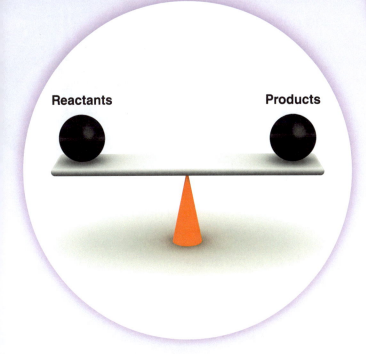

Reactants   Products

Mass is also conserved in **chemical reactions**. That means that the reactants (the things at the start of the reaction which are reacting) will have the same mass as the products that are formed.

1. If a candle of mass 564g is lit, and is allowed to melt completely to form liquid wax, what is the mass of the liquid wax?

2. What particles are found in the nucleus of an atom?

3. What is the charge of an electron?

4. Explain the difference between an element and a compound.

KEYWORDS

Solute (noun) ➤ The solid that dissolves in the solvent to form a solution.

Solvent (noun) ➤ The liquid that the solute dissolves in.

Pure substance (noun) ➤ Contains only one type of atom or molecule.

Impure substance (noun) ➤ Contains more than one type of atom or molecule.

A **pure substance** contains only one type of atom or molecule, e.g. a sample of pure water contains only H_2O molecules. An **impure substance** has impurities so it contains more than one type of atom or molecule, e.g. sea water contains other molecules and ions as well as H_2O molecules.

Mixtures

➤ A mixture contains two or more substances that are mixed, but not chemically bonded.

➤ Solutions are a type of mixture. A solution is formed when a **solute** is dissolved into a **solvent**.

➤ Solutions have the same mass as the **solute** and the **solvent** put together.

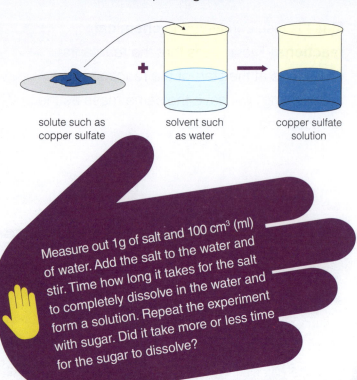

solute such as copper sulfate + solvent such as water → copper sulfate solution

Measure out 1g of salt and 100 cm³ (ml) of water. Add the salt to the water and stir. Time how long it takes for the salt to completely dissolve in the water and form a solution. Repeat the experiment with sugar. Did it take more or less time for the sugar to dissolve?

Diffusion

➤ Diffusion occurs in liquids and gases.

➤ In diffusion, particles move from a high concentration (more particles) to an area of low concentration (less particles).

➤ Diffusion occurs faster in gases than in liquids, gas particles in a gas move faster.

Methods of separation

As the substances which make up a mixture are not chemically bonded, mixtures can be separated by physical processes. There are a variety of different methods that can be used to separate mixtures.

Filtration

➤ Mixtures of solids and liquids can be separated using filtration.

➤ A mixture is poured through a filter and the liquid passes through pores (small holes) in the filter. The solid will be left on the filter and the liquid will pass through and can be collected.

➤ The size of the filter's pores will determine the size of the solid particles which are filtered out.

➤ Filtration will not filter out the **solute** from a solution.

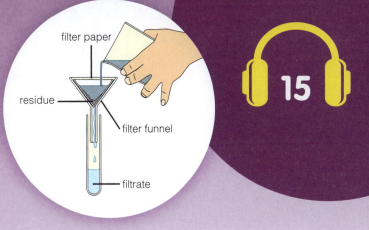

Evaporation

➤ **Solutes** can be separated from **solvents** by evaporation.

➤ The solution is left in an open container. The solvent will evaporate to form a gas. The solid solute will be left behind.

➤ Evaporation can be speeded up by heating the solution.

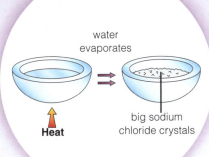

water evaporates

big sodium chloride crystals

Heat

Chromatography

➤ Chromatography is used to separate mixtures of different coloured dyes or a mixture of substances which have different solubilities in the solvent being used.

➤ A sample of dye mixture is placed on filter paper, to which a **solvent** is then added.

➤ As the **solvent** moves across the filter paper, it carries the dyes with it.

➤ As the dyes have different solubilities, they will move different distances across the filter paper.

solvent front

starting point

X   Ink A   Ink B   Ink C   Ink D

Distillation

➤ Distillation can be used to separate a solvent from a solution or a mixture of liquids with different boiling points.

➤ The solution is heated to the boiling point of the solvent.

➤ The **solvent** forms a gas, which is then cooled by a condenser.

➤ When the gas is cooled it forms a pure liquid.

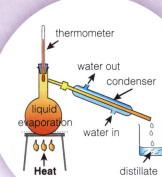

thermometer

water out

condenser

liquid evaporation

water in

Heat

distillate

Identification of pure substances

➤ Melting and boiling points can be used to determine the purity of a substance.

➤ Impurities cause the boiling and melting points of a substance to change.

➤ Impurities generally lower the melting point and raise the boiling point.

1. In what direction will particles move by diffusion?

2. What happens to the melting point of a substance that contains impurities?

3. When a solid dissolves in a liquid, what term is given to the liquid?

4. When a solid dissolves in a liquid, what term is given to the solid?

5. What determines the size of the particles that pass through a filter?

16

Chemical reactions involve the rearrangement of atoms to form new molecules.

Chemical equations

Chemical reactions can be represented by **word equations**:

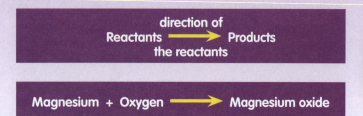

direction of
Reactants ➡ Products
the reactants

Magnesium + Oxygen ➡ Magnesium oxide

They can also be represented by **symbol equations**. In symbol equations the chemical symbols for atoms and compounds are used. A large number in front of the symbol indicates how many molecules there are. A subscript number (e.g. $_2$) shows how many atoms of an element there are in a molecule:

$$2Mg + O_2 \longrightarrow 2MgO$$

As mass is conserved, the products must have the same number of atoms as the reactants.

Types of chemical reactions

There are different types of chemical reaction:

Combustion

In a combustion reaction a substance is burned in air. This causes it to react with oxygen:

Carbon + Oxygen ➡ Carbon dioxide
$$C + O_2 \longrightarrow CO_2$$

Thermal decomposition

In a thermal decomposition reaction a chemical breaks down (decomposes) when heated. An example is shown below:

Calcium carbonate ➡ Calcium oxide + Carbon dioxide
$$CaCO_3 \longrightarrow CaO + CO_2$$

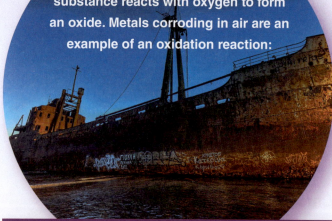

Oxidation
In an oxidation reaction a substance reacts with oxygen to form an oxide. Metals corroding in air are an example of an oxidation reaction:

Aluminium + Oxygen ➡ Aluminium oxide
$$4Al + 3O_2 \longrightarrow 2Al_2O_3$$

Displacement

In a displacement reaction a more reactive element will displace (take the place of) a less reactive one. In the example below, potassium (K) is more reactive than sodium (Na). Potassium therefore displaces sodium to form potassium chloride:

Potassium + Sodium chloride ➡ Potassium chloride + Sodium
$$K + NaCl \longrightarrow KCl + Na$$

Acids and alkalis

The strength of an **acid**, or **alkali**, can be determined using the pH scale. Indicators are used to show pH value and the universal indicator changes colour in different pHs.

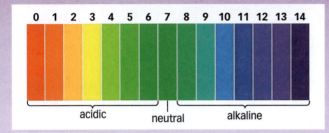

- ➤ Strong acids have low pH
- ➤ Water is neutral and has a pH of 7
- ➤ Strong alkali have high pH

The indicator phenolphthalein is colourless in acidic solutions and **pink** in alkaline solutions.

Strong acids and alkalis are corrosive. This means they can cause burns to skin and will damage other material too.

When acids are added to alkalis they will neutralise each other. The neutralisation effect depends on how strong the acid and alkali are.

- ➤ Weak acid + Weak alkali ⟶ Neutral
- ➤ Strong acid + Weak alkali ⟶ Weak acid
- ➤ Weak acid + Strong alkali ⟶ Weak alkali
- ➤ Strong acid + Strong alkali ⟶ Neutral

KEYWORDS

Reactants (noun) ➤ The substance which reacts in a chemical reaction.
Products (noun) ➤ The substance which is produced by a chemical reaction.
Acid (noun) ➤ A chemical which has a low pH (below pH 7)
Alkali (noun) ➤ A chemical which has a high pH (above pH 7)

Catalysts

A catalyst is a substance which speeds up the rate of chemical reaction. A platinum catalyst is used in car exhaust systems to remove the toxic products produced by the burning of the fuel in the engine. Enzymes are biological catalysts.

On separate cards write the word 'combustion', a short explanation of what a combustion reaction is and an equation of a combustion reaction. Repeat for oxidation, displacement, metal + acid and acid + alkali. Now mix all the cards up and try and match each reaction to the correct description and example.

Acid + Alkali ⟶ Salt + Water

Hydrochloric acid + Sodium hydroxide ⟶ Sodium chloride + Water
$HCl + NaOH ⟶ NaCl + H_2O$

Acid + Metal ⟶ Salt + Hydrogen

Hydrochloric acid + Magnesium ⟶ Magnesium chloride + Hydrogen
$2HCl + Mg ⟶ MgCl_2 + H_2$

1. What pH is neutral?
2. What is produced when acids react with metals?
3. What is produced when an acid reacts with alkalis?
4. In what type of reaction is a substance burnt in air?
5. In what type of reaction is a substance reacted with oxygen to form an oxide?

MODULE 17
ENERGETICS

During changes of state of matter, energy is either taken in or released by a substance.

Heating

When a solid is heated it absorbs energy and its temperature will rise until it reaches its melting point.

The solid will then melt to form a liquid. Whilst the solid is melting its temperature will stay the same even though the solid is still absorbing energy.

Once it has become liquid, if it continues to be heated, it will absorb further energy and its temperature will rise.

Cooling

When a gas cools down its temperature decreases and it releases energy.

When a gas reaches its boiling point its temperature will remain constant as it condenses to form a liquid. Once the liquid has formed, and it continues to be cooled, its temperature will continue to drop as it releases energy.

Endothermic reactions

In an **endothermic reaction** energy is taken in from the environment during the reaction. This means that the temperature of the reaction will decrease. An example of an endothermic reaction is the reaction between ethanoic acid and sodium carbonate.

Take some ice and place it in a cup, use a thermometer to record the ice's temperature every five minutes until it has completely melted. Plot a graph of your results.

Exothermic reactions

In an **exothermic reaction** energy is released into its environment. This means that the temperature of the reaction will increase. Neutralisation reactions between acids and alkalis are examples of exothermic reactions.

Endothermic reaction (noun) ➤ A reaction which gains energy from its environment and takes energy in.

Exothermic reaction (noun) ➤ A reaction which transfers energy to its environment and releases energy.

Endothermic:	Ethanoic acid + Sodium carbonate ⟶ Sodium ethanoate + Carbon dioxide + Water
Exothermic:	Nitric acid + Sodium hydroxide ⟶ Sodium nitrate + Water HNO_3 + $NaOH$ ⟶ $NaNO_3$ + H_2O

Once the liquid reaches its boiling point, it will boil and become a gas.

As the liquid is boiling, it will continue to absorb energy, but again its temperature will not rise. Once it has boiled and formed a gas, if it continues to be heated, its temperature will once again rise.

The temperature stays the same during these changes of state as the energy is being used to break the attractions between the molecules.

When the liquid reaches its melting point, it will still release energy and the liquid will become a solid.

At this point its temperature will remain constant. Once it becomes a solid and it continues to cool, its temperature will drop as it releases energy.

1. What term is given to a reaction which transfers energy to the environment?
2. What term is given to a reaction which absorbs energy from the environment?
3. What will happen to the temperature of a solid that is melting?
4. When a very hot object is cooling, why does its temperature decrease?
5. What will happen to the temperature of a liquid as it boils?

MODULE 18
THE PERIODIC TABLE

The groups are numbered 1 to 7 from left to right. The last group on the right is group 0 or 8.

All elements are arranged in the **periodic table** in order of their atomic number. The first version of the modern periodic table was constructed by Dmitri Mendeleev in 1869.

The periodic table can be used to determine the relative reactivity of an element.

➤ The elements become more reactive further down groups 1 and 2 (potassium is more reactive than lithium).

➤ The elements become less reactive further down group 7 (fluorine is more reactive than bromine).

In the periodic table the horizontal rows are the **periods**.

The vertical columns are the **groups**.

Elements in the same group have similar properties.

The non-metal elements are found on the right-hand side of the periodic table.

Metals vs non-metals

The metals make up the majority of the periodic table. Elements between groups 2 and 3 are the transition metals.

➤ All metals are solid at room temperature, except for mercury.

➤ Eleven non-metals are gases at room temperature, e.g. hydrogen and nitrogen.

➤ One non-metal is a liquid (bromine).

➤ All the other non-metals are solids at room temperature, e.g. carbon

Metals	Non-metals
Shiny when freshly cut.	Dull (not shiny).
Malleable (they can be bent without breaking).	Brittle (break or shatter easily).
Good conductors of heat and electricity.	Insulators (poor conductors of heat and electricity).
Hard and strong.	They are weak.
Most metals have a high density.	Most non-metals have a low density.
Sonorous (they make a ringing sound when they are struck).	Not sonorous (do not produce a ringing sound when struck).

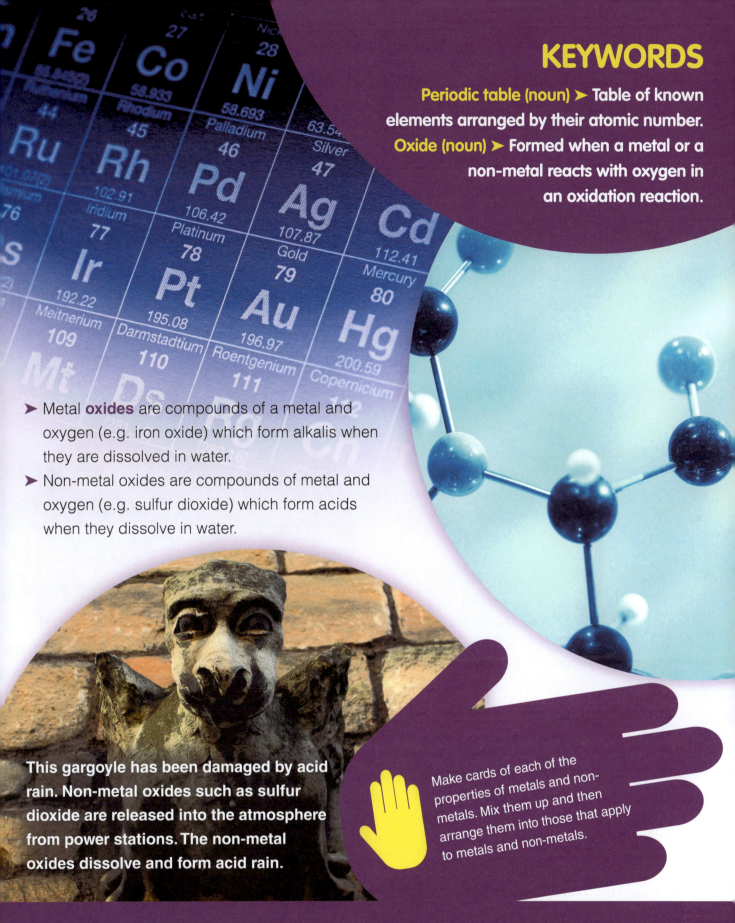

KEYWORDS

Periodic table (noun) ➤ Table of known elements arranged by their atomic number.
Oxide (noun) ➤ Formed when a metal or a non-metal reacts with oxygen in an oxidation reaction.

➤ Metal **oxides** are compounds of a metal and oxygen (e.g. iron oxide) which form alkalis when they are dissolved in water.

➤ Non-metal oxides are compounds of metal and oxygen (e.g. sulfur dioxide) which form acids when they dissolve in water.

This gargoyle has been damaged by acid rain. Non-metal oxides such as sulfur dioxide are released into the atmosphere from power stations. The non-metal oxides dissolve and form acid rain.

Make cards of each of the properties of metals and non-metals. Mix them up and then arrange them into those that apply to metals and non-metals.

1. What forms when a metal oxide dissolves in water?
2. Are non-metals generally poor conductors or good conductors of heat?
3. Are all metals solids at room temperature?
4. Most non-metals are liquids at room temperature. True or false?
5. What name is given to the vertical columns of the periodic table?

MODULE 19
MATERIALS

Metal elements can be placed in a **reactivity series**, starting with the most reactive metal and decreasing in reactivity. Reactivity refers to how readily a metal will react with other elements. Carbon and hydrogen are non-metals, but they are often included in the reactivity series.

Most reactive

potassium K
sodium Na
calcium Ca
magnesium Mg
carbon C
zinc Zn
iron Fe
lead Pb
hydrogen H
copper Cu
gold Au

Least reactive

Iron oxide + Carbon $\longrightarrow$ Iron + Carbon dioxide

$2Fe_2O_3 + 3C \longrightarrow 4Fe + 3CO_2$

Extraction of metals

➤ Unreactive metals, such as silver, gold and platinum, can be dug out of the ground as pure metals. Other metals react to form rocks called ores.

➤ In order to obtain a metal it must be extracted from its ore.

➤ Metals that are less reactive than carbon can be extracted using a **displacement reaction**, by heating the ore with carbon.

➤ Iron is an example of a metal extracted in this way.

KEYWORDS

Displacement reaction (noun) ➤ A reaction where a more reactive element displaces a less reactive element from a compound.

Reactivity series (noun) ➤ Metals, carbon and hydrogen, placed in order of reactivity, with the most reactive metal first.

Ceramics

Made from crystalline compounds.

Have a wide range of uses, including pots, plates, cups and brake discs on cars.

Ceramics are hard, brittle (will break and shatter rather than bend) and are good thermal and electrical insulators (they are poor conductors).

Polymers

The properties of polymers will depend on the monomers they are made from.

Very large molecules made by joining lots of smaller molecules (monomers) together.

Polypropene (many propene monomers joined together) is used in packaging.

Polyethene (many ethene monomers joined together) is used to make plastic bags.

Composites

Composites are made from two or more different materials.

An example of a composite material is carbon fibre, which is extremely strong and light, and is used to make Formula 1 cars.

When these materials are combined they form a composite that has properties that are different from the original materials.

Write each of the metals in the reactivity series (and carbon and hydrogen) on separate pieces of card. Put the metals in order of reactivity, starting with the most reactive.

1. Which metal is more reactive – aluminium or gold?

2. Is iron more reactive than carbon?

3. In what form is iron found in rock?

4. What is a polymer?

5. What type of material is carbon fibre?

MODULE 20
EARTH AND ATMOSPHERE

The Earth is mainly composed of the elements iron, oxygen, silicon, magnesium, sulfur, nickel, calcium and aluminium. There are also traces (very small amounts) of other elements.

The atmosphere

The Earth's atmosphere is composed the following gases:

➤ Nitrogen 78%
➤ Oxygen 21%
➤ Argon 0.9%
➤ Carbon dioxide 0.04%
➤ Other trace gases 0.06%

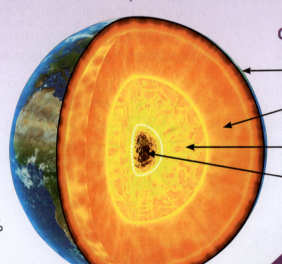

Composition of the Earth

crust – the outer thin layer of solid rock which we live on

mantle – a very large layer of rock. The mantle is solid, but can move over very long time scales

outer core – liquid rock and it is extremely hot

inner core – the hottest part of the planet. It is solid and made up of iron and nickel

Create cards out of each of the stages of the rock cycle and the carbon cycle. Mix the cards up and then arrange the two cycles into the correct order.

Rock type	Formation	Examples
Igneous rock	Molten (liquid) rock which cools. Extrusive igneous rocks are formed from lava, cooling quickly above the ground to form small crystals. Intrusive igneous rocks are formed from magma, cooling slowly underground to form large crystals.	Basalt – extrusive Granite – intrusive
Sedimentary rock	Layers of sediment which build up over long periods of time are compacted and heated. Fossils are found in sedimentary rocks.	Sandstone, limestone
Metamorphic rock	Sedimentary, or igneous rocks, are forced underground and are heated to very high temperatures and are put under extreme pressure, forming a different rock. Fossils can be found in metamorphic rocks but are deformed.	Marble, slate

Humans are producing carbon dioxide by:
➤ Burning fossil fuels.
➤ Cutting down and then burning trees.

As trees take in carbon dioxide in photosynthesis, cutting down trees in deforestation contributes to an overall increase of atmospheric carbon dioxide.

The carbon cycle shows the flow of carbon between the atmosphere and the surface of the Earth.

Burning fossil fuels as fuel (combustion) releases carbon dioxide.

Over very long periods of time the remains of dead animals and plants can form fossil fuels such as coal, oil and gas.

Animals and plants produce waste and die. They are broken down by decomposers who also respire, releasing more carbon dioxide into the atmosphere.

Animals and plants respire, releasing carbon dioxide back into the atmosphere.

Organic molecules of plants are fed on by animals.

The carbon is used to form organic molecules.

Carbon dioxide is absorbed by plants by photosynthesis.

KEYWORDS

Igneous rock (noun) ➤ Formed when molten rock cools.
Sedimentary rock (noun) ➤ Formed from sediment.
Metamorphic rock (noun) ➤ Formed when sedimentary rock is under extreme heat and pressure.

➤ All the resources humans require are obtained from the Earth and most will eventually run out.
➤ Recycling is extremely important.

Carbon dioxide is a greenhouse gas.
▼
The greenhouse gases trap thermal energy in the atmosphere.
▼
There is an overall rise in average global temperatures.
▼
This can lead to climate change and an increase in the frequency of extreme weather.
▼
Some parts of the world will get colder whilst other areas will become warmer.

1. What layer of rock is found beneath the crust?
2. Why are the crystals in intrusive igneous rocks large?
3. What rocks are formed when sedimentary rocks are forced underground and heated under high pressure?
4. What is the most common gas in the atmosphere?
5. What environmental problem is caused by high concentrations of carbon dioxide in the atmosphere?

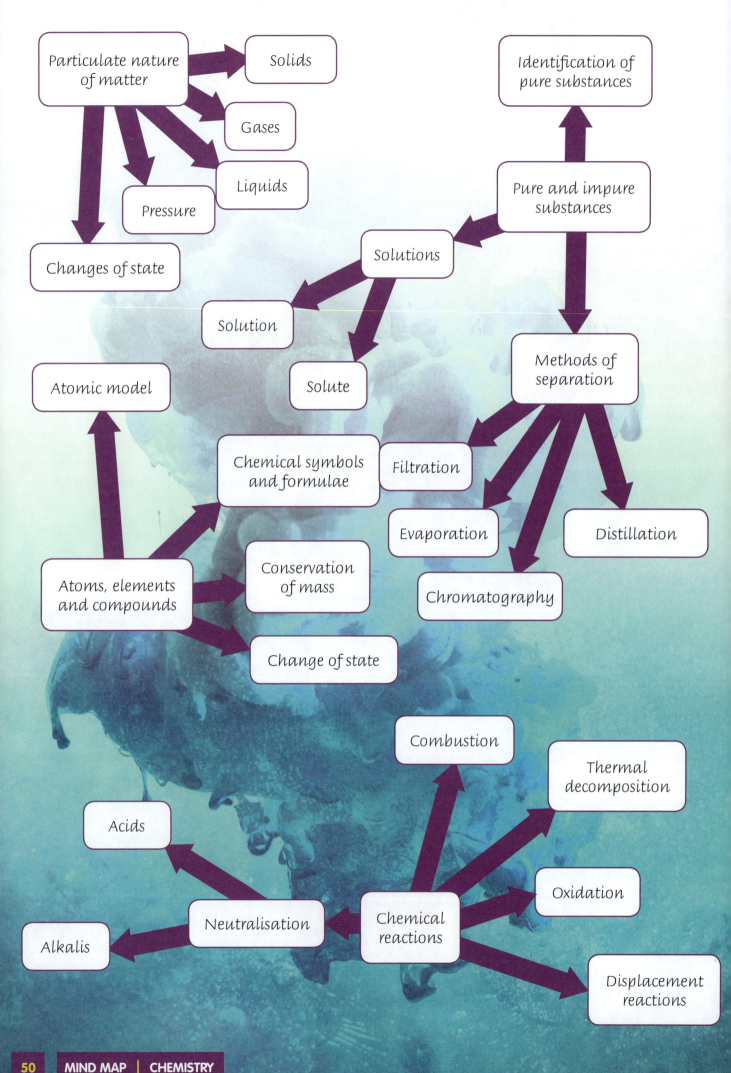

Particulate nature of matter

Solids

Gases

Liquids

Pressure

Changes of state

Atomic model

Chemical symbols and formulae

Atoms, elements and compounds

Conservation of mass

Change of state

Solutions

Solution

Solute

Identification of pure substances

Pure and impure substances

Methods of separation

Filtration

Evaporation

Distillation

Chromatography

Combustion

Thermal decomposition

Acids

Oxidation

Neutralisation

Chemical reactions

Alkalis

Displacement reactions

Patterns in reactions

Using carbon to obtain metal from its oxides

The periodic table

Reactivity series

Non metals

Materials

Ceramics

Metals

Polymers

Metal oxides

Non-metal oxides

Composites

Igneous

Sedimentary

Metamorphic

Composition and structure of the Earth

Production of carbon dioxide by humans

Rock cycle

Carbon cycle

Exothermic

Earth and atmosphere

The atmosphere

Recycling

Heating

Endothermic

Cooling

Energetics

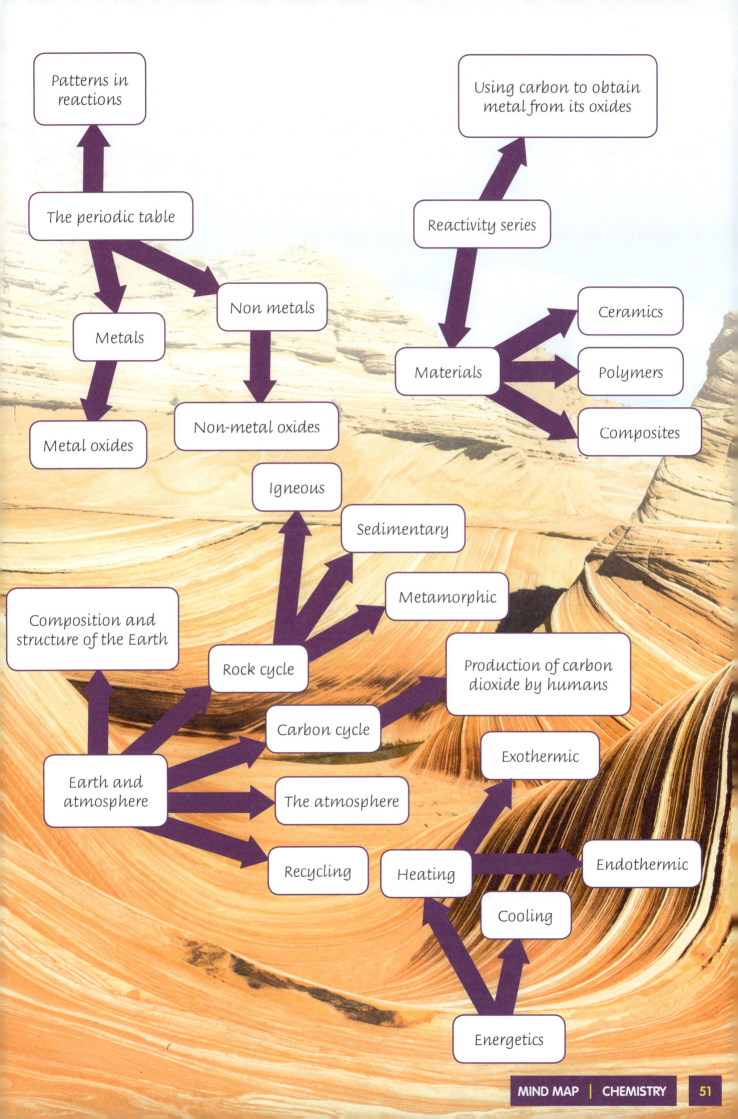

CHEMISTRY
PRACTICE QUESTIONS

1. **a)** Match the following descriptions to the types of rock. (3)

A.	May contain fossils which aren't deformed.		Sedimentary
B.	Formed from other rocks deep underground by high pressure and temperatures.		Igneous
C.	Can be formed when lava cools on the earth's surface.		Metamorphic

Tony and Jasmine were examining some samples of igneous rock that had large crystals.

Tony said he thought that they were intrusive igneous rocks and the crystals were large because they had cooled quickly.

Jasmine thought that they were intrusive rocks, but the crystals were large because they had formed slowly.

b) Who was correct? (1)

c) Explain your answer. (2)

2. An unknown element reacted with a series of different metal oxides. The results are shown below:

	Iron oxide	Copper oxide	Aluminium oxides
Displacement reaction occurs?	No	Yes	No

a) Complete the table below to show which metals are more reactive than A and which are less reactive than A. (2)

More reactive than A	Less reactive than A

b) Could carbon be used to extract metal A from its ore? (1)

c) Explain your answer. (2)

3. As part of a practical investigation Becky has dissolved a solid in water.

   a) What term is given to the solid that has dissolved? (1)

   b) What term is given to the water that the solid has dissolved in? (1)

   Becky now wants to separate the solid from the solution. Steven suggests using the equipment shown below.

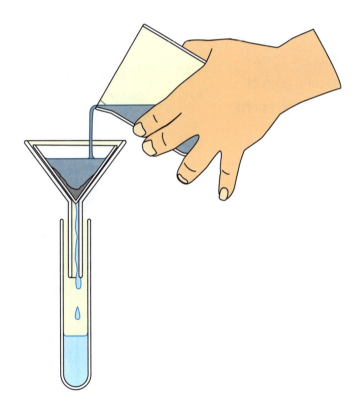

   c) Name the separation process this equipment will carry out? (1)

   d) Explain why this equipment can't be used to separate the solid from the solution. (3)

   e) Name one separation technique Becky could use to separate the solid from the water. (1)

MODULE 21
CALCULATION OF FUEL COSTS IN THE DOMESTIC CONTEXT

21

Different foods contain different amounts of energy. Energy in food is measured in kilojoules (kJ).

Energy in foods

The energy in packaged food is shown on its label. It's important to take note of the energy, as eating too many high energy foods can lead to a person becoming overweight and suffering from obesity.

Power

Power is the energy transferred over time. It is measured in watts (W). All electrical appliances have a power rating, showing how much energy they transfer over time. A fridge's power rating may be 500W, whilst a laptop's may be 50W.

Energy transferred

The energy transferred in an hour is a kW hour (kWh). Electricity bills for a home are calculated per kWh.

Here is an example fuel bill showing how much energy a family used:

Time period	Energy used
Jan 2013–March 2013	600 kWh

This family pays **15p** per kWh. What is this family's fuel bill for the time period shown above?

0.15 × 600 = £90

Energy transferred from fuels

Fossil fuels, e.g. coal, oil and gas, are produced over millions of years from dead organisms

Fossil fuels are burnt in power stations

The energy is used to heat water to make steam

The steam then turns a turbine which generates electricity

Fossil fuels are **non-renewable** sources of energy. They have taken millions of years to produce and so they can not be replaced in a lifetime. The burning of fossil fuels also produces carbon dioxide, which contributes to the **greenhouse effect**, which is causing **climate change**.

Nuclear power uses radioactive elements to heat water rather than burning fossil fuels. Nuclear power doesn't increase the concentration of carbon dioxide, in the atmosphere, but it does produce highly dangerous **radioactive waste**.

Renewable energy sources will not run out, they include:

➤ **Wind power** turns turbines to generate electricity.

➤ **Solar power** uses light and heat energy from the sun to generate electricity.

➤ **Wave power** is used to generate electricity.

➤ The rise and falling of the **tides** is used to generate electricity.

➤ **Geothermal power** from hot rocks beneath the surface of the Earth, is used to heat water to produce steam, which is then used to turn turbines and generate electricity.

➤ **Biomass** is the burning of wood to heat water, which produces steam to turn a turbine to generate electricity. When the trees used to produce biomass are grown, they take in carbon dioxide. Therefore, when using biomass as a fuel, there is no overall (net) increase in the concentration of carbon dioxide in the atmosphere.

KEYWORDS

Power (noun) ➤ Energy transferred over time.

Renewable energy (noun) ➤ An energy resource that will not run out, for example, wind power.

Non–renewable energy (noun) ➤ An energy resource that will run out, for example, fossil fuels.

Advantages of renewable energy	Disadvantages of renewable energy
Do not require a fuel which will eventually run out.	Do not provide a constant supply of electricity, e.g. wind turbines only generate electricity when the wind is blowing.
Do not increase the concentration of carbon dioxide in the atmosphere, so they do not contribute to climate change.	Building renewable energy facilities often means destroying important natural habitats, e.g. tidal power stations can lead to the destruction of coastal wetlands.
Large start up costs, but more cost effective over time.	Can only be sited in certain locations, e.g. geothermal requires hot rocks near to the surface of the Earth and solar power relies on getting enough sunshine.

Research the energy contained in different food types from food labels. Which foods contain the largest amounts of energy?

1. What is power?

2. What unit represents the energy transferred in an hour and is used when calculating electricity bills?

3. Give two examples of fossil fuels.

4. What would be the outcome of eating foods that contained too much energy?

5. Give one disadvantage of using wind power?

MODULE 22
ENERGY CHANGES AND TRANSFERS

When there is a temperature difference between two objects, there will be an energy transfer from the hotter object to the colder object until both objects are the same temperature (in **equilibrium**).

➤ If the two objects are touching, the thermal energy will be transmitted by **conduction**.
➤ If the objects are not touching, the thermal energy will be transferred by **radiation**. Radiation is the method by which the Sun's heat is transmitted to the Earth.

To reduce the loss of heat energy an **insulator** can be used. In winter, houses are warmer than the surrounding air due to them being heated. This means that the house will lose heat to the surrounding air. To reduce this loss of heat homeowners can fit:

➤ double glazing
➤ loft insulation
➤ cavity wall insulation.

All these methods will reduce the heat that is lost to the air from a house.

Type of energy	Description
Chemical energy	Contained in fuels and food.
Elastic potential energy	Energy in a stretched spring.
Gravitational potential energy	Energy in an object that is above the surface of the Earth.
Sound energy	Sound
Light energy	Light
Electrical energy	Current flowing through a circuit.
Nuclear energy	Released by radioactive elements.
Thermal energy	Heat

Some examples of energy transfers include:
➤ Burning fuel – the **chemical energy** in the fuel is converted into **light energy** and **thermal** energy.
➤ Stretching a spring – the **kinetic energy** is converted into **elastic potential energy**. When the spring is released the **elastic potential energy** will be converted into **kinetic energy**.

When an object is dropped the object's **gravitational potential energy** is converted into **kinetic energy** as it falls.

When food is metabolised the **chemical energy** in the food is converted into different forms, including **kinetic energy** and **thermal energy**.

When an electrical circuit is completed the **chemical energy** in the battery is converted to **electrical energy** in the circuit.

KEYWORDS

Conduction (noun) ➤ The transfer of thermal energy between objects which are in contact.

Radiation (noun) ➤ The transfer of thermal energy that doesn't require a medium and can occur through a vacuum.

Insulator (noun) ➤ A material which is a poor conductor of heat and can be used to reduce heat loss from houses.

Machines can be used to produce a **force**.
➤ Some machines can produce a big force. However, they are unable to produce small, fine movements.
➤ Some machines produce a small force. A machine that produces a smaller force can produce finer movements.

Go around your house and identify objects which have the following types of energy: kinetic, chemical, elastic potential, gravitational potential, thermal, sound and light. What energy transfers occur when these objects are used?

1. How is heat transmitted from the Sun to the Earth?
2. If two objects are touching how will heat be transferred between them?
3. What energy transfer occurs when a ball is dropped from a height above the ground?
4. Why is it important to fit insulation in a house?
5. What energy transfer occurs when a battery is wired into a circuit?

MODULE 23
CHANGES IN SYSTEMS

Energy cannot be created or destroyed. It can only be transferred from one form to another. Therefore the energy at the beginning of a process is the same as the energy at the end of the process.

When a fuel such as coal is burned, the stored chemical energy is converted to heat energy, light energy and energy in the product (carbon dioxide).

Coal (15kJ/g) ➤ Heat + Light + Carbon Dioxide
= 15KJ/g

KEYWORD

Conservation of energy (noun)
➤ Energy cannot be created or destroyed; it is transferred from one form to another.

In an **exothermic** reaction (one which releases energy to the environment) the products will have a lower energy than the reactants.

In an **endothermic** reaction (a reaction which takes in energy from the environment) the products will have a higher energy than the reactants.

Draw a comic strip to illustrate the energy changes in this module. Make sure you show what form the energy is being transformed to.

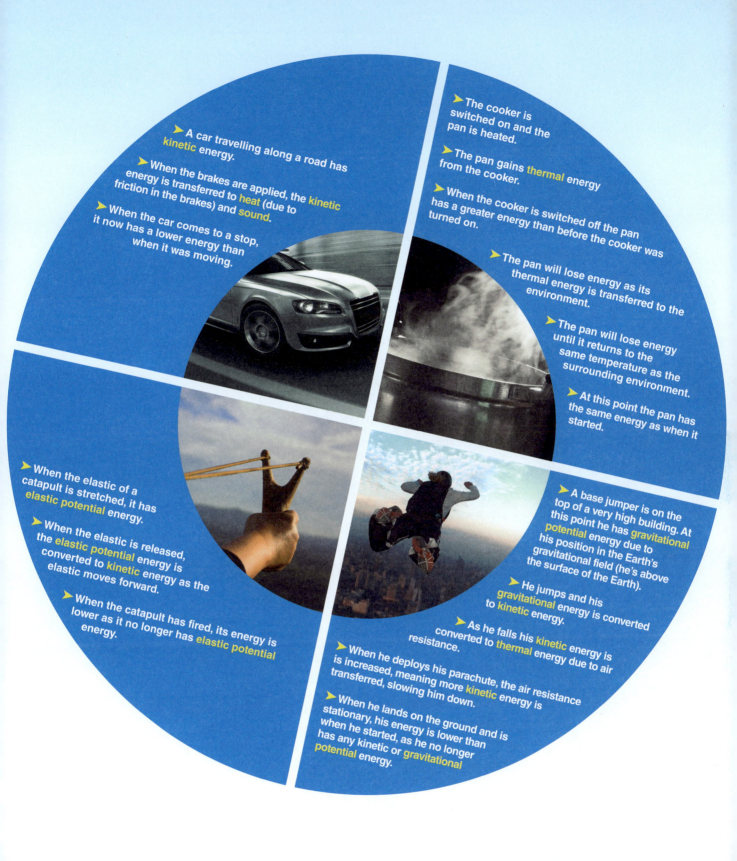

A car travelling along a road has **kinetic** energy.

When the brakes are applied, the **kinetic** energy is transferred to **heat** (due to friction in the brakes) and **sound**.

When the car comes to a stop, it now has a lower energy than when it was moving.

The cooker is switched on and the pan is heated.

The pan gains **thermal** energy from the cooker.

When the cooker is switched off the pan has a greater energy than before the cooker was turned on.

The pan will lose energy as its thermal energy is transferred to the environment.

The pan will lose energy until it returns to the same temperature as the surrounding environment.

At this point the pan has the same energy as when it started.

When the elastic of a catapult is stretched, it has **elastic potential** energy.

When the elastic is released, the **elastic potential** energy is converted to **kinetic** energy as the elastic moves forward.

When the catapult has fired, its energy is lower as it no longer has **elastic potential** energy.

A base jumper is on the top of a very high building. At this point he has **gravitational potential** energy due to his position in the Earth's gravitational field (he's above the surface of the Earth).

He jumps and his **gravitational** energy is converted to **kinetic** energy.

As he falls his **kinetic** energy is converted to **thermal** energy due to air resistance.

When he deploys his parachute, the air resistance is increased, meaning more **kinetic** energy is transferred, slowing him down.

When he lands on the ground and is stationary, his energy is lower than when he started, as he no longer has any kinetic or **gravitational potential** energy.

1. What energy does a fuel such as coal contain?

2. When the coal is burnt, in what form is the energy released?

3. How could a base jumper gain gravitational potential energy?

4. When an electric kettle is switched off it has low energy. When it is turned on it gains energy. Why does the kettle gain energy?

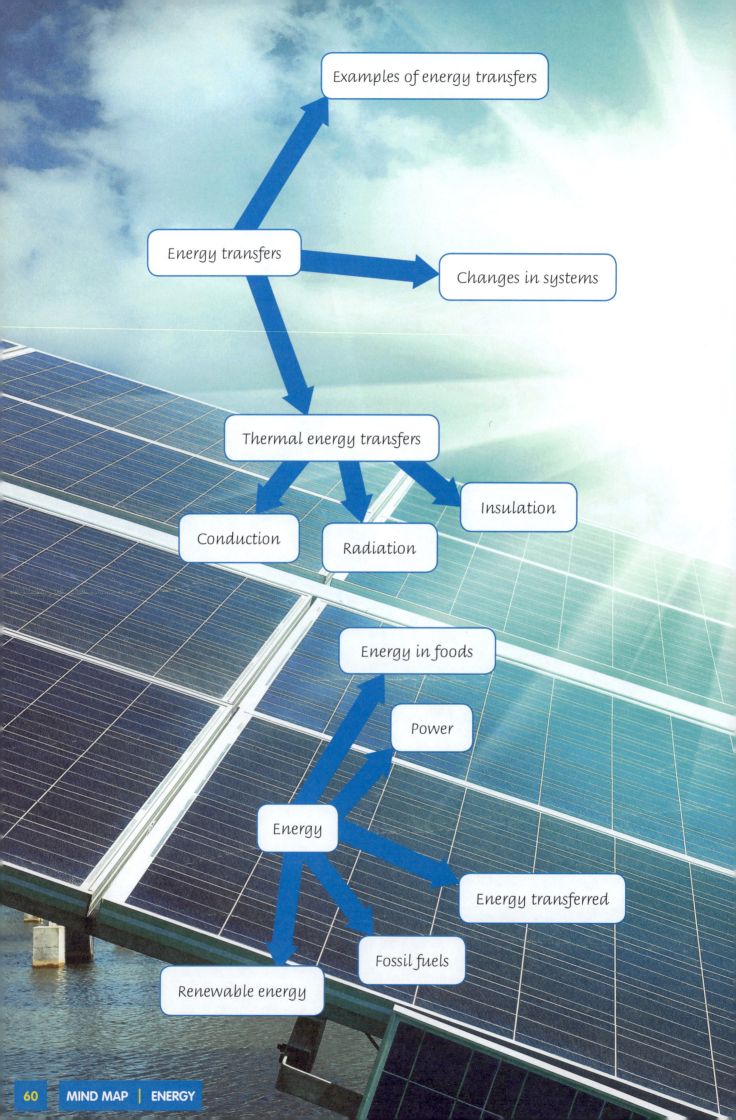

Examples of energy transfers

Energy transfers

Changes in systems

Thermal energy transfers

Conduction

Radiation

Insulation

Energy in foods

Power

Energy

Energy transferred

Fossil fuels

Renewable energy

ENERGY
PRACTICE QUESTIONS

1. Below is the energy bill for a family for nine months.

Time period	Energy used (kWh)
Jan – Mar	800
Apr – Jun	687
July – Sep	452

  a) During which period did the family use the most energy? (1)

  b) The family pays 16p per kWh. How much did the family spend on energy over this nine-month period? (1)

  c) The family would like to fit insulation in their house. What effect would this have on the family's energy bill? (1)

  d) Explain why this effect would occur. (2)

2. Environmental groups would like to reduce the use of fossil fuels around the world.

  a) What environmental problems are caused by burning fossil fuels? (1)

The environmental campaigners would like to switch to using renewable sources of energy instead of non-renewable fossil fuels.

  b) Why are fossil fuels non-renewable? (1)

  c) Give two examples of renewable energy resources. (2)

  d) What would be the problem with relying only on wind power for electricity generation? (2)

MODULE 24
DESCRIBING MOTION

24

An object's **speed** can be calculated using the distance it covers and the time it takes to cover that distance.

Units for speed vary, but they will always include distance and time. Metres per second (m/s) is one of the most commonly used units: distance is measured in metres and time is measured in seconds.

Speed (m/s) = Distance (m) ÷ Time (s)

A graph of distance plotted against time can be drawn to show a journey. The graph below shows the journey of a motorcyclist:

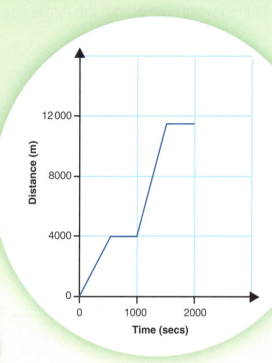

A car travels 500m in 36 seconds. What is its speed?

500 ÷ 36 = 13.9 m/s

The equation can be rearranged to allow distance or time to be calculated:

Distance = Speed × Time
Time = Distance ÷ Speed

➤ From 0–500 seconds (s) the bike is travelling away from the start point so there is a steep, upward line.
➤ From 500–1000s the motorcyclist has stopped, so the line is horizontal.
➤ At 1000s the bike sets off again. The line is steeper than it was between 0–500s. This shows the bike is travelling faster than it was between 0–500s.
➤ The bike stops again between 1500s and 2000s.

KEYWORDS

Speed (noun) ➤ Distance travelled over time.

Relative motion (noun) ➤ Change in position of an object relative to another object.

Relative motion

➤ A car is travelling alongside a train. Both the car and the train are travelling at the same speed. To a passenger travelling on the train it would look like the car was staying relatively in the same place.

➤ The train then begins to speed up and is now travelling faster than the car.

➤ As the train speeds up it would appear to the passenger on the train that the car was moving backwards.

Measure out a short distance and roll a ball. Time how long it takes the ball to travel the distance you've measured.

Use the formula **speed = distance ÷ time** to calculate the object's average speed.

Repeat the investigation, but this time push the ball faster. Calculate the new speed. How much faster was the second object? Draw distance time graphs for both objects. How do they differ?

1. What is the formula for calculating speed?

2. A sprinter runs 100m in 9.82 seconds. What is his speed?

3. A bird flies at a speed of 12 m/s. How long will it take for the bird to travel 750m?

4. How far will an arrow travelling at 60 m/s fly in 1.5 seconds?

When objects come into contact they can exert push or pull **forces** on each other. The push and pull forces on objects can be represented by arrows. Force can be measured in units called, **Newtons**.

The size of the forces on an object determines the effect. If the forces on an object are the same size and are acting in opposite directions, the forces are balanced. If the forces are different sizes and/or are not acting in opposite directions, the forces are unbalanced.

When any object moves through a medium, forces act against movement.

➤ Moving through the air leads to air resistance. The faster the movement the greater the air resistance.

➤ Water also resists movement.

➤ When travelling along a surface, friction occurs between the two surfaces, acting against the direction of movement.

	Moving object	**Stationary object**
Forces balanced	The object will continue to move at the same speed in the same direction.	The object will remain stationary.
Forces unbalanced	The object will either speed up or slow down and may change direction.	The object will begin to move in the direction of the greatest force.

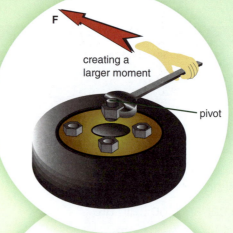

creating a larger moment

pivot

creating a small moment

pivot

Forces can cause objects to turn around pivots. The turning force around a pivot is a **moment**.

As the car's speed increases, the friction opposing the car's movement also increases:

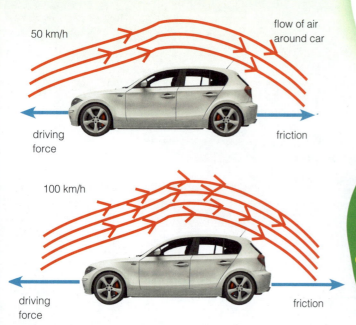

50 km/h

flow of air around car

driving force

friction

100 km/h

driving force

friction

On a large sheet of paper draw diagrams to show the effects of:

➤ balanced forces;

➤ unbalanced forces in opposite directions;

➤ unbalanced forces in different direction on both a stationary and a moving object.

You can choose the object, for example, a car, a plane, a boat etc.

MODULE 25
FORCES, MOTION AND BALANCED FORCES

25

When a force is applied to an object, the object exerts an opposing force. If these forces are balanced, the object will remain unchanged. An example is a person standing on the Earth.

Forces can even act when objects don't touch. Examples of these non-contact forces are gravity, magnetism and static electricity.

KEYWORD

Newton (noun) ➤
A unit of force.

When a weight is added to a spring the force of the weight is greater than the force of the spring so the spring is stretched.

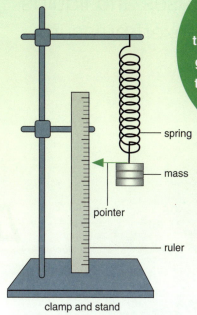

spring

mass

pointer

ruler

clamp and stand

If a person's weight is 750N, the Earth pushes up with an equal force, meaning the person remains standing on the Earth.

➤ When a push force on an object is greater than the opposing force exerted by the object, the object will be compressed.

➤ When a pull force on an object is greater than the opposing force of the object, the object will be stretched.

➤ Hooke's Law states that the force needed to extend or compress a spring is proportional to the distance that it is being extended or compressed.

1. When the forces on a stationary object are unbalanced what will happen to the object?

2. When the forces on a moving object are balanced what will happen to the object?

3. When moving through the air what force opposes the movement?

4. Give two examples of non-contact forces.

5. What are units of force?

MODULE 26
PRESSURE IN FLUIDS

Pressure is the ratio of force over the area the force is applied. Pressure acts at the normal (i.e. at right angles) to the surface. Gases and liquids are fluids and exert pressure.

Pressure

➤ Pressure can be increased by increasing the force applied, or decreasing the area over which the force is applied.

➤ Pressure can be decreased by decreasing the force applied, or increasing the area over which the force is applied.

If a boat is less dense than the water it's floating on, its weight will be balanced by upthrust

If a boat is more dense than the water, its weight will not be balanced by upthrust and it will sink

Upthrust is a force that opposes the weight of an object placed in a fluid. An object that is less dense than water will float on the surface of the water. The weight of the object is balanced by the upthrust.

An object which is denser than the water will sink. The weight of the object is greater than the upthrust, so the forces are unbalanced.

KEYWORDS

Atmospheric pressure (noun) ➤ Pressure exerted by gases in the atmosphere. As you go up through the atmosphere the atmospheric pressure decreases.

Upthrust (noun) ➤ A force that opposes the weight of an object in a fluid.

Atmospheric pressure

Air in the atmosphere exerts pressure (atmospheric pressure).

At sea level the standard atmospheric pressure is 1 bar.

Travelling upwards decreases the atmospheric pressure.

At higher altitudes there is less air above to exert pressure.

Pressure in liquids

Liquids exert pressure.

The pressure increases with depth.

As depth increases there is more liquid above therefore a greater force is exerted.

The deepest point in the ocean is the Mariana Trench in the Pacific Ocean, where there is 10km of water above the bottom of the trench. This exerts a pressure of over 1000 bar, which is over a thousand times the pressure at sea level.

The plane is at high altitude so the air pressure is lower as there are fewer air particles above it

Plane

The balloon is at low altitude so the air pressure is still high as there are a lot of air particles above it

Balloon

Earth

Design and build a small boat that will float in a sink. Think carefully about the materials you can use. Remember the boat must be less dense than the water in order for the upthrust to allow it to float.

1. What happens to the pressure on an object that is sinking through water?
2. An object which is more dense than water will float on the surface of the water. True or false?
3. In which direction does pressure act on a surface?
4. What force opposes the weight of an object in water?
5. What is pressure?

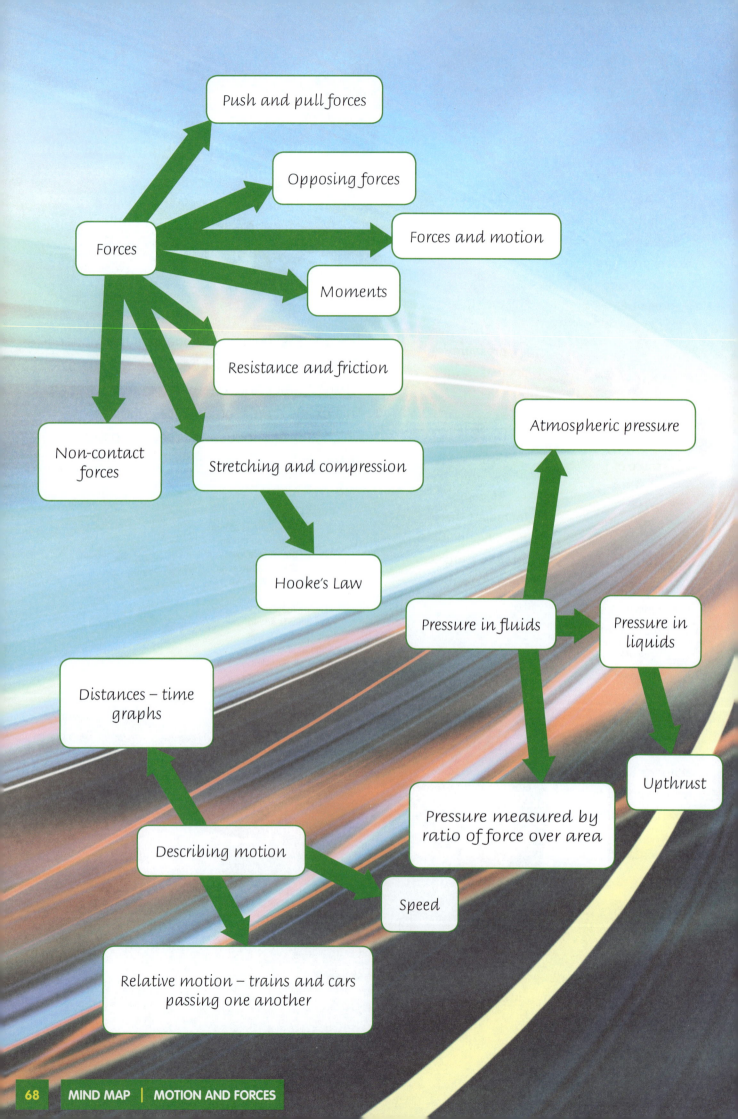

Push and pull forces

Opposing forces

Forces and motion

Forces

Moments

Resistance and friction

Atmospheric pressure

Non-contact forces

Stretching and compression

Hooke's Law

Pressure in fluids

Pressure in liquids

Distances – time graphs

Upthrust

Describing motion

Pressure measured by ratio of force over area

Speed

Relative motion – trains and cars passing one another

MOTION AND FORCES
PRACTICE QUESTIONS

1. **a)** A ship is sailing through water at a constant speed. What statement can be made about the forces on the ship? (1)

   b) Why is it important to consider upthrust when designing a ship? (2)

   c) A ship deploys a submarine to investigate marine life. The submarine has to be built to withstand very high pressures. Explain why this is important. (3)

2. A jet ski travels 510 metres in 30 seconds.

   a) Calculate the speed of the jet ski. (1)

   Speed = Distance ÷ Time

   b) The jet ski slows down to 9 m/s. How long will it take for the jet ski to travel 100m? (1)

   Time = Distance ÷ Speed

   The graph below shows the movements of the jet ski over 30 minutes.

   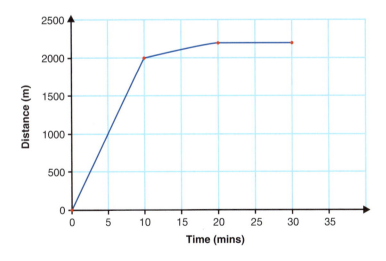

   c) During which period is the jet ski (i) moving fastest and (ii) stationary? (2)

MODULE 27
OBSERVED WAVES, SOUND WAVES AND ENERGY AND WAVES

Water waves are transverse waves and sound waves are longitudinal waves.

Observed waves

➤ Water waves are undulations that travel through the water.
➤ Water waves have a **transverse** motion through water. Transverse motion means that a particle on the wave moves up and down as the wave passes.
➤ Waves can be **reflected**.
➤ When two waves hit each other they can either add together or cancel each other out. This is known as **superposition**.

direction of sound wave   vibration of sound wave

Sound waves

Sound travels as a longitudinal wave. Sound needs to travel through a medium (a solid, a liquid or a gas). Sound can't travel through outer space because space is a vacuum.

Sound travels at different speeds through different mediums:

➤ Speed of sound in the air = 343.2 metres per second (m/s).
➤ Speed of sound in water (at 20°C) = 1,481 m/s.
➤ Speed of sound in a solid varies depending on the solid, for example = 6420 m/s in aluminium and 3962 m/s in glass.

Frequency

➤ The number of waves that pass a point in one second.
➤ Measured in Hz.
➤ A low pitched sound will have a low **frequency**.
➤ A high pitched sound will have a high frequency.

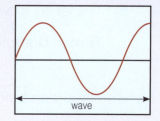

wave

low pitched sound

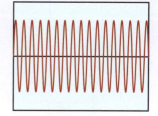

high pitched sound

Reflection

When a sound wave reflects off a surface, the sound can be heard again (an echo). A flat, smooth surface will reflect sound, whilst a bumpy surface will absorb sound.

'This foam is used to sound-proof rooms so they can be used as music studios'

Sound is produced by the vibration of objects. In a loud speaker, the cone of the speaker vibrates to produce the sound.

Microphone

Human ear

A microphone detects sound when the sound wave hits the diaphragm of the microphone.

This causes the diaphragm to vibrate.

The vibrations are converted into an electrical signal.

The sound waves cause the eardrum in the ear to vibrate.

The vibrations of the eardrum cause tiny bones in the ear to move.

These bones push fluid inside the ear.

The movement of the fluid is detected by tiny hairs.

These tiny hairs send an electrical nerve impulse to the brain.

Ultrasound

Pressure waves can transfer energy. **Ultrasound** is an example of a pressure wave. Ultrasound is sound which has a higher frequency than the auditory range of humans. Ultrasound has many different applications:

➤ Ultrasound can be used to clean delicate objects, e.g. jewellery, lenses and surgical tools.

➤ Ultrasound can be used in physiotherapy to treat muscle injuries. It stimulates blood circulation and cell activity. This reduces pain and speeds up healing.

Organism	Auditory Range
Humans	20Hz up to 20 kHz
Dogs	40 Hz to 60 kHz
Bats	10 Hz up to 200 kHz

Write each of the keywords on this page on a small card. On separate cards write the definition for each word. Mix the cards up and try to match the definition to the keyword.

27

1. What term describes the number of waves which pass a point in a second?
2. What type of wave is a sound wave?
3. Which animals can hear the highest frequency sounds, humans or dogs?
4. What type of surface will best reflect sound?
5. Does sound travel fastest in a solid, a liquid or a gas?

KEYWORDS

Transparent (adjective) ➤ Allows light to pass through.

Opaque (adjective) ➤ Does not allow light to pass through.

Refraction (noun) ➤ Light bends when it travels between media with different densities.

The angle of reflection is equal to the **angle of incidence** (the angle of the incoming light). Both these angles are measured from the **normal** (90° to the surface of the mirror).

Light does not require a medium to travel through and can travel through a vacuum.

➤ The speed of light in a vacuum is 299 792 458m/s. Light is the fastest thing in the universe.

➤ Light travels in straight lines, so its behaviour can be represented by a ray diagram.

Diffuse reflection

In diffuse reflection, light is reflected from a surface in many different directions. This means that an image is not formed. Diffuse reflection occurs from most objects and allows us to see them.

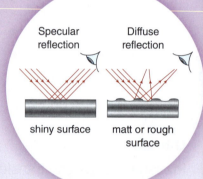

Specular reflection — shiny surface

Diffuse reflection — matt or rough surface

Light travels through **transparent** materials such as glass. Light cannot travel through **opaque** materials. When light strikes an opaque material it is **reflected**.

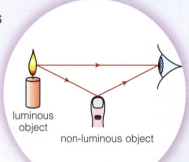

luminous object

non-luminous object

Not all light may be reflected by an object, because some light may be **absorbed**. The frequencies of light which are absorbed determine what colour we perceive an object to be.

➤ Red objects reflect red light into our eyes and absorb all other colours of light.

➤ Black objects absorb all light, so no light is reflected into our eyes.

➤ White objects reflect all colours of light.

➤ Light waves have a frequency and the frequency determines the colour of the light.

➤ A prism can be used to split white light into different wavelengths.

➤ Each colour of light refracts slightly differently and this produces the spectrum of colours.

➤ The spectrum is always made up of the same colours in the same order - red, orange, yellow, green, blue, indigo, violet.

Specular reflection

In specular reflection, light is reflected in a single direction. As the light is reflected an image is formed. When you look in a mirror the fact that you see your reflection is due to specular reflection.

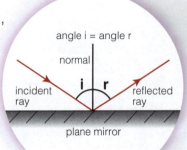

angle i = angle r

normal

incident ray

i r

reflected ray

plane mirror

When light passes from one medium to another (for example, from a gas to a solid) it bends. This process is known as **refraction**.

➤ When light moves into a less dense medium, it speeds up and bends away from the normal.
➤ When light moves into a more dense medium, it slows down and bends towards the normal.

When light passes through a **convex lens**, the rays of light are refracted. Convex lenses can be used to focus light on a point.

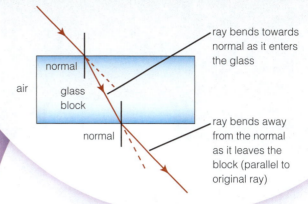

ray bends towards normal as it enters the glass

ray bends away from the normal as it leaves the block (parallel to original ray)

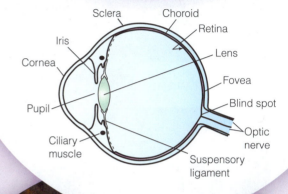

➤ Digital cameras have a light-sensitive layer which absorbs light coming into the camera.
➤ The light is converted it into electrical signals.
➤ The camera uses the electrical signals to form an image.

➤ The human eye is designed to take in light and focus it on the **retina**.
➤ The **lens** at the front of the eye can change shape to focus the light on the retina.
➤ When the light hits the retina, it causes a chemical reaction in the cells of the retina, producing an electrical signal that is transmitted along the **optic nerve**.
➤ The optic nerve carries the electrical signal to the brain where it's interpreted and we perceive what we are seeing.

➤ Light rays hit the photographic film.
➤ A chemical reaction occurs.
➤ An image is produced on the film.

On separate cards, write 'specular reflection', 'diffuse reflection', 'refraction', and definitions of each of these key terms and diagrams to represent them. Mix the cards up and then try to match the terms to the diagrams and descriptions

1. Can light travel through a vacuum?
2. When light passes from one medium to another what happens?
3. What will happen when white light is shone through a prism?
4. What type of reflection occurs when a person looks into a mirror and sees their reflection?
5. What colours of light are absorbed by a blue object?

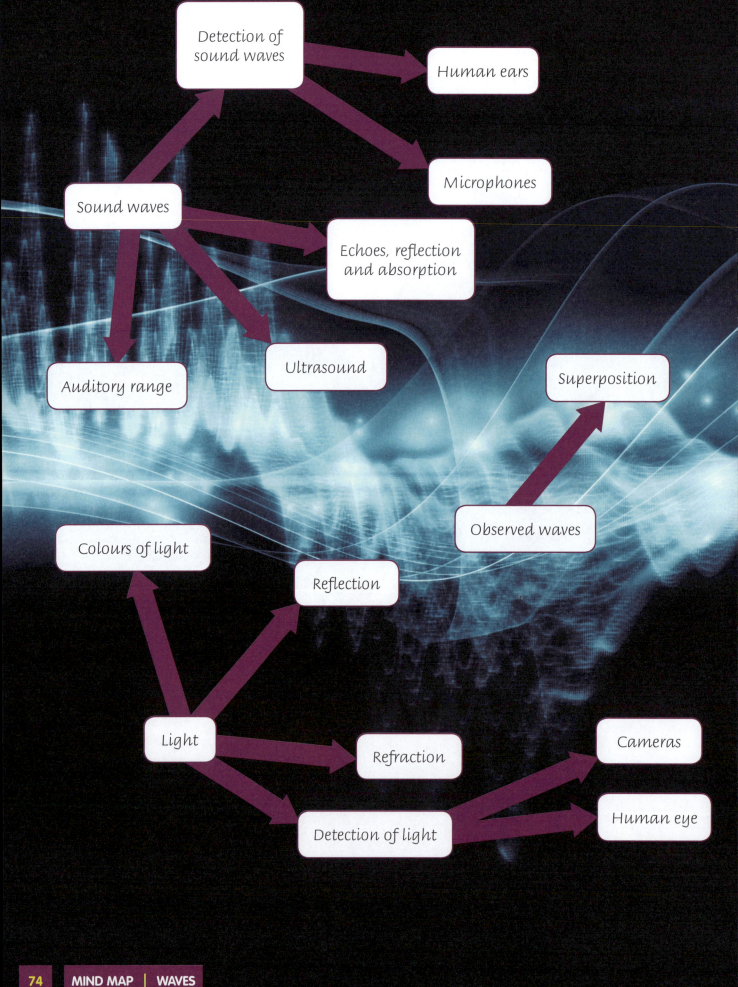

Detection of
sound waves

Human ears

Microphones

Sound waves

Echoes, reflection
and absorption

Ultrasound

Auditory range

Superposition

Observed waves

Colours of light

Reflection

Light

Cameras

Refraction

Human eye

Detection of light

WAVES
PRACTICE QUESTIONS

1. **a)** Paul has a dog whistle. When he blows it he hears no sound, but his dog does. Explain why this is. (2)

   b) Paul's dog jumps in the water to chase some ducks. What type of waves does he produce when he jumps in? (1)

   c) Superposition occurs to these waves. Which of the below statements best describes superposition? (1)

   A. Superposition is when waves overlap. They will always cancel each other out.

   B. Superposition is when waves overlap. They will always add together.

   C. Superposition is when waves overlap. They could add together or cancel each other out.

   D. Superposition is when waves do not come into contact with each other.

2. **a)** During a fireworks display, the fireworks are seen to explode before the sound is heard. Why is this? (1)

   b) Harry is standing a distance away from a large wall. After each firework explodes he hears an echo of the bang. Which of the following statements best explains this? (1)

   A. The sound is refracted off the wall.

   B. The sound is reflected off the wall.

   C. The sound travels faster through the solid wall.

   D. The echo is due to diffuse scattering of light.

   c) If a firework was detonated in a vacuum, the explosion would be seen, but there would be no sound. Why is this? (2)

MODULE 29
CURRENT ELECTRICITY

Electricity powers electrical appliances. **Electrical current** is the flow of **charge** through conductors such as wires and electrical components. Charge can be made to flow by using a cell or a battery.

In order for an electric current to flow there must be a complete circuit and it must form a loop from one side of the cell or battery to the other. Wires connect the cell or battery to the components to form a **circuit**. Diagrams of a complete and incomplete circuit are shown to the right:

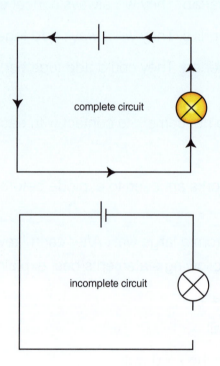

Electrical circuits can be wired in **series** or **parallel**. In a series circuit, the components are arranged in a line, whilst a parallel circuit has branches:

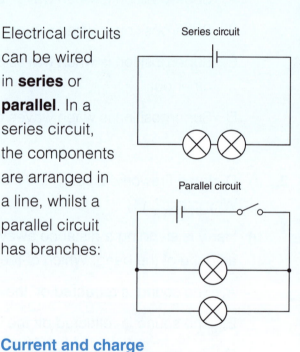

Current and charge

Current is a measure of how much charge is flowing in a circuit. Current is measured in **amperes** (A) using an **ammeter**.

➤ In a **series circuit**, the current is the same at all points in the circuit.
➤ In a **parallel circuit**, the current splits between the branches. This means that the current can be different at different points in the circuit. When two branches meet the current travelling, each of the branches adds together.

The wires in a circuit diagram are shown as straight lines and the components are represented by symbols:

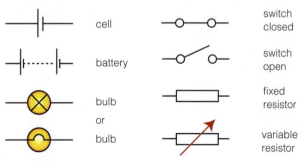

Potential difference

Potential difference is a measure of the difference in energy between two parts of a circuit. Potential is measured in Volts (V) using a voltmeter.

Batteries and bulbs have ratings showing the potential differences they are designed to work at, for example AA batteries have a rating of around 1.5V and car headlight bulbs have a rating around 12V.

Resistance

All conductors resist the flow of charge through them. Resistance is the ratio of potential difference to current and is measured in **ohms**.

Resistance of components

➤ A conducting component has a low resistance. This means it will allow a large current to flow through it.

➤ An insulating component has a high resistance. This means it will only allow a small current to flow through it.

For example:

➤ A wire is a conducting component. It is a good conductor with a low resistance.

➤ A resistor is an insulating component. It has a higher resistance, so limits the current that can flow through the circuit.

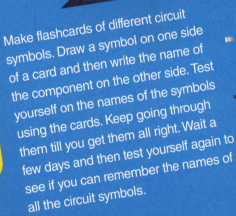

KEYWORDS

Electrical current (noun) ➤ The flow of charge through a circuit.

Series circuit (noun) ➤ A circuit where all the components are wired in a single loop.

Parallel circuit (noun) ➤ A circuit that has branches.

Potential difference (noun) ➤ The difference in energy between two parts of a circuit.

Make flashcards of different circuit symbols. Draw a symbol on one side of a card and then write the name of the component on the other side. Test yourself on the names of the symbols using the cards. Keep going through them till you get them all right. Wait a few days and then test yourself again to see if you can remember the names of all the circuit symbols.

1. What are the units of resistance?
2. What are the units of voltage?
3. Would an insulating component have a low or a high resistance?
4. The current in a parallel circuit is the same at all points in the circuit. True or false?
5. What is electrical current?

MODULE 30
STATIC ELECTRICITY AND MAGNETISM

Magnetism and the forces due to static electricity are examples of non-contact forces.

Static electricity

When two insulators are rubbed together, electrons are transferred from one insulator to another.

Electrons are particles with a **negative charge**. When they are transferred from one object to another it becomes **negatively charged**. The object which is losing the electrons will become **positively charged**. This is called **static electricity**.

electrons

Two positively charged objects will repel each other (be pushed away). Two negatively charged objects will also repel each other. A positively charged object will be attracted to a negatively charged object.

Electric field

➤ Electric fields are generated by electrically charged particles and magnetic fields.
➤ Electric fields will cause objects which are not in contact with each other to exert forces on each other (be attracted or repelled).

Magnetism

Magnets attract **magnetic** materials such as iron, steel, cobalt and nickel. Magnets do not attract **non-magnetic** materials such as carbon, aluminium and plastic.

The strongest parts of a magnet are its two poles, **north** and **south**. The Earth's magnetic North Pole is located in the Arctic. If a bar magnet is suspended, it will align itself so it is pointed towards the magnetic North Pole.

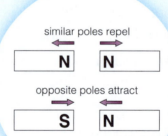

similar poles repel

opposite poles attract

Compasses have a magnetic needle that is attracted to the Earth's magnetic North Pole. This allows compasses to be used for navigation.

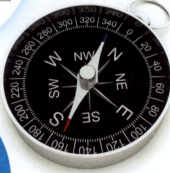

Magnetic fields

Magnets produce a **magnetic field** where the magnetic effects are felt.

Magnetic fields can be plotted using a compass. The compass will point along the **field lines** produced by a magnet. By moving the compass around the magnet, it's possible to plot the field lines. The field lines produced by a bar magnet are shown opposite.

Electromagnet

➤ When a current is passed through a wire, it produces a magnetic field.

➤ By wrapping a wire in a coil around an iron core, the magnetic field can be magnified to form an electromagnet.

➤ Electromagnets are very useful, because the magnetic field is only generated when a current is flowing through the wire. This means that an electromagnet can be turned on or off.

➤ The strength of the magnetic field can be increased by increasing the number of coils of the wire, or increasing the current flowing through the wire.

➤ The magnetic field of an electromagnet is the same shape as that of a bar magnet.

A **D.C. Motor** can be made by suspending a coil of wire between the opposite poles of two magnets. When an electric current is passed through the wire, the wire will experience a force and turn.

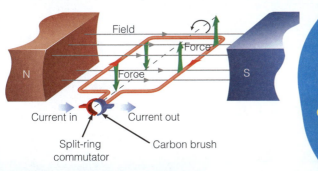

Field · Force · Force · N · S · Current in · Current out · Split-ring commutator · Carbon brush

KEYWORD

Magnetic (adjective) ➤ **A material which is attracted to a magnet.**

Electromagnets are used to pick up scrap metal. When the current to the magnet is switched off, the metal will drop.

Inflate a balloon and rub it with a cloth or against a jumper. The balloon will become charged. If you bring the balloon close to your hair (or someone else's), you'll see it attract the hair. You can extend this investigation by tearing up a piece of paper into very small pieces. The charged balloon should attract these pieces and pick them up if you bring the balloon into close contact with them.

1. What type of motor is made by suspending a coil of wire between the opposite poles of two magnets?

2. What happens when the north pole of one magnet is brought close to the south pole of another magnet?

3. Give one way that the strength of an electromagnet can be increased.

4. What will be the effect of stopping the current flowing through the wire of an electromagnet?

5. What is the shape of the magnetic field of an electromagnet?

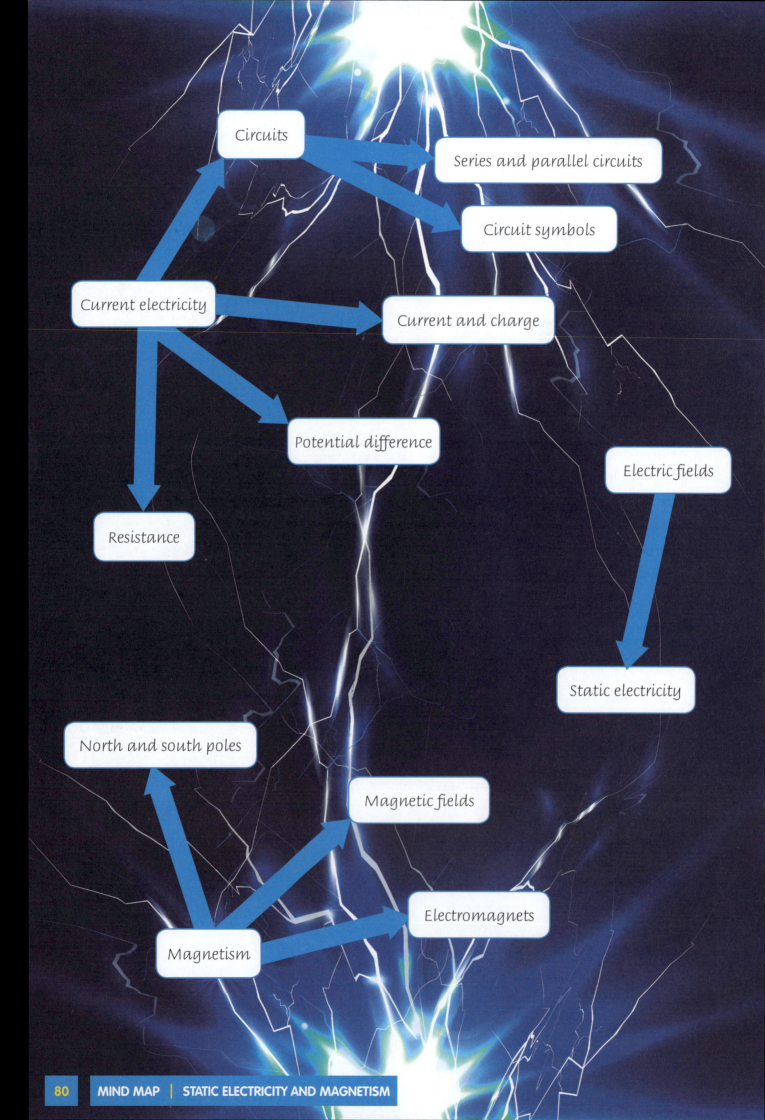

Circuits

Series and parallel circuits

Circuit symbols

Current electricity

Current and charge

Potential difference

Electric fields

Resistance

Static electricity

North and south poles

Magnetic fields

Electromagnets

Magnetism

ELECTRICITY AND ELECTROMAGNETISM
PRACTICE QUESTIONS

1. a) Mike and Caroline are investigating static electricity, Mike rubs a perspex rod with a cloth. When he brings the rod close to Caroline's hair, it attracts it.

Which of the following statements best explains this effect? (1)

    A. The perspex rod and Caroline's hair are both negatively charged, so they repel each other.

    B. The perspex rod and Caroline's hair are both positively charged, so they attract each other.

    C. The perspex rod and Caroline's hair have opposite charges, so they repel each other.

    D. The perspex rod and Caroline's hair have opposite charges, so they attract each other.

b) Complete the statements below: (2)

The rod becomes _____ charged because it loses electrons.

Electrons are _____ charged.

c) Mike and Caroline go onto investigate magnetism. Mike knows that the when electricity is passed through a wire it produces a magnetic field. How can this effect be magnified to produce an electromagnet? (1)

d) What piece of equipment could be used to determine the shape of the field lines of a magnet? (1)

e) What shape are the field lines produced by an electromagnet? (1)

2. a) Identify the two circuits shown below. (2)

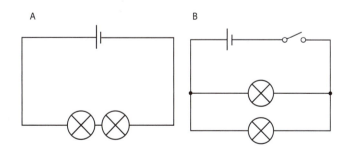

b) Match the terms below to their correct units (3)

a. potential difference		i.	Ohms
b. current		ii.	Volts
c. resistance		iii.	Amperes

MODULE 31
PHYSICAL CHANGES

31

Matter can undergo many physical changes. In all of these changes the matter is conserved. When matter is converted into another state, the total mass of the matter remains the same.

Process	Change of state
Melting	Solid to liquid
Freezing	Liquid to solid
Evaporation	Liquid to gas
Sublimation	Solid directly to gas
Condensation	Gas to liquid
Dissolving	A solute dissolves into a solvent to form a solution

➤ All solids, liquids and gases have a mass.
➤ Solids and liquids have a definite volume. The volume of a gas can be changed.
➤ Solids have a definite shape. The shape of liquids and gases is determined by the container they are stored in.

Density is a measure of the mass of an object in a given volume. Solids are denser than liquids, whilst liquids are denser than gases. As solids and liquids are relatively **incompressible**, they have a fixed density.

Gases are **compressible**, so their density can be altered. The more gas particles there are in a set volume, the higher the density and the higher the pressure, e.g. gas is compressed in a fire extinguisher.

Brownian motion is where gas particles move randomly.

➤ Diffusion occurs where gas and liquid particles are able to move.

➤ Diffusion is the movement of particles from an area with a high concentration to an area with a lower concentration.

➤ Diffusion will occur until the concentrations are equal and equilibrium is reached.

➤ In a chemical change, atoms are rearranged to form new compounds and molecules.

➤ In a physical change, the atoms of the matter stay the same, but the physical state of the matter changes.

KEYWORDS

Density (noun) ➤ The measure of a mass in a given volume.

Brownian motion (noun) ➤ Brownian motion is the random movement of particles which are suspended in a fluid (a liquid or a gas). Brownian motion occurs because the liquid and gas particles randomly collide with the suspended particle, e.g. smoke particles moving through the air.

Write the words 'melting', 'freezing', 'evaporation', 'sublimation', 'condensation' and 'dissolving' on to small cards. Write an explanation of each of these terms onto different cards. Mix the cards up and try and match the correct terms to the explanations.

1. What is sublimation?
2. What is dissolving?
3. What states of matter shows Brownian motion?
4. How can the density of a gas be increased?
5. In what direction does diffusion occur?

MODULE 32
PARTICLE MODEL AND ENERGY IN MATTER

When particles are heated they gain kinetic energy. This means they move faster and are able to overcome the bonds holding them together.

When a **solid** is heated its particles gain energy.

When it reaches its melting point, the particles are able to move relative to each other and are no longer as tightly packed together.

The matter can now flow and it no longer has a fixed shape.

As the particles aren't as closely packed together, the density is also reduced.

The solid has become a liquid.

When a **liquid** is heated to its boiling point, the particles have enough kinetic energy so that they are able to move away from each other.

The matter will fill any container, as it no longer has a fixed volume.

As the particles are able to move far away from each other, the density of the matter also decreases. The liquid has become a gas.

The reverse happens as the gas cools to form a liquid and the liquid cools to form a solid. Liquid water cooling to form ice is an exception.

- Liquid water is more dense than solid ice. This is why ice floats on liquid water.
- Ice is less dense than water because the bonds between the water molecules force them apart. This means that the molecules are further apart in ice than they are in liquid water.

Fuels contain stored chemical energy. When fuels are burnt (combustion) the chemical energy is converted into thermal energy and light energy.

Organic compounds that organisms use as food also contain stored chemical energy. The energy is released by respiration.

Write out each of the stages of the flow charts on cards. Mix the cards up and then arrange them to show the correct sequence of events of how a solid becomes a gas when heated.

1. What state of matter does not have a fixed volume?
2. Are the particles closer together in a gas or a liquid?
3. In what state is water densest?
4. What type of stored energy does food contain?

MODULE 33
SPACE PHYSICS

Gravity attracts all objects to each other. As the strength of gravity is proportional to the mass of the object, we only feel the force of gravity from objects with a very large mass, such as planets and stars.

The Earth's gravity holds objects to its surface. **Weight** is the force caused by gravity. Weight can be calculated using the following formula:

Weight = Mass × Gravitational field strength (g)

On Earth the gravitational field strength is 10N/kg (g = 10N/kg). For example, a person has a mass of 84kg:

84 × 10 = 840N

The gravitational field strength is different on different planets. Jupiter has a much larger mass than the Earth, so it has a much higher gravitational field strength. The same person would have a much larger weight on Jupiter, as shown in the calculation above.

840N

A man has a mass of 84kg on Jupiter where the gravitational field strength is 25N/kg:

84 × 25 = 2100N

➤ The Earth's gravity keeps the moon in orbit around the Earth.
➤ The Sun has a very large mass and so its gravity is very strong. The Sun's gravity keeps the Earth and the other planets in orbit around the Sun.

Space

The Sun is a star which generates heat and light by nuclear fusion. The Earth and the other planets orbiting the Sun make up the **solar system**. Our sun is one of 300 billion stars in our galaxy, the Milky Way. There are over 100 billion galaxies in the universe.

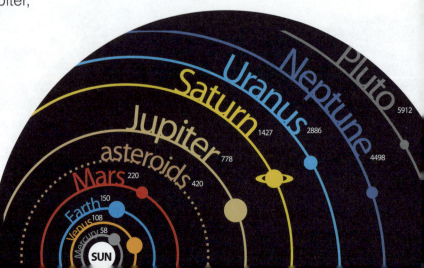

Neptune 4498
Pluto 5912
Uranus 2886
Saturn 1427
Jupiter 778
asteroids
Mars 220
420
Earth 150
Venus 108
Mercury 58
SUN

Weight (noun) ➤ The force caused by gravitational attraction.

Outside the solar system, distances are so huge that a special unit of distance, **light years**, is used to measure them. A light year is the distance travelled by light in one year (9500 billion km).

➤ Proxima Centauri, the closest star to the Sun, is 4.24 light years away.
➤ Canis Major Dwarf Galaxy, the closest galaxy to the Sun, is 25,000 light years away.

The seasons

The **seasons** are caused by the tilt of the Earth as it orbits. This means that at different times of the year, either the northern half of the Earth (the **northern hemisphere**) or the southern half (the **southern hemisphere**) is more directly exposed to energy from the Sun. This leads to the different day lengths and temperatures of the seasons.

During winter – the hemisphere is tilted away from the Sun: days are shorter, nights are longer and temperatures are lower.

During summer – the hemisphere is tilted towards the Sun: days are longer, nights are shorter and temperatures are higher.

When it is summer in the northern hemisphere, its winter in the southern hemisphere; when it's winter in the northern hemisphere it's summer in the southern hemisphere.

Using paper and string construct a hanging mobile showing the Sun, Earth and Moon and their relative positions. You can extend this activity by adding in the other planets in the solar system.

1. What is a light year?
2. In what season are days shorter in the northern hemisphere?
3. Put the following in size order from smallest to largest: planet, galaxy, solar system, universe.
4. When it is winter in the northern hemisphere, what season will it be in the southern hemisphere?
5. Why does the Sun have a greater gravitational attraction than the Earth?

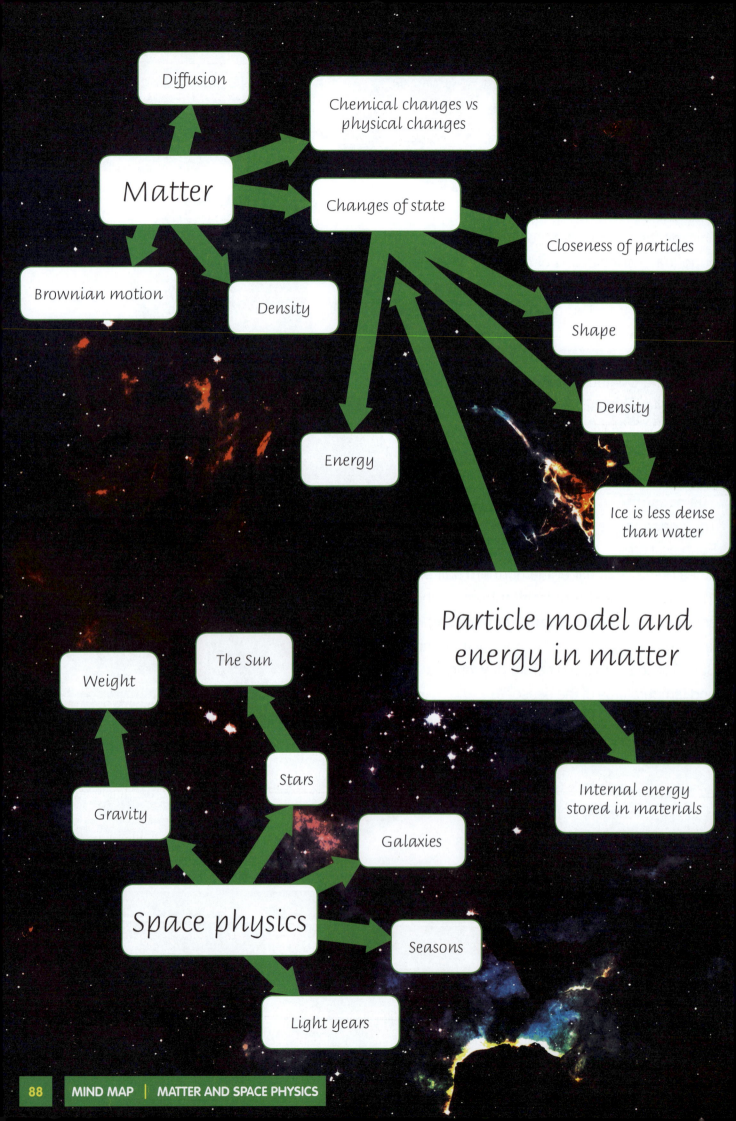

Diffusion

Chemical changes vs physical changes

Matter

Changes of state

Closeness of particles

Brownian motion

Density

Shape

Energy

Density

Ice is less dense than water

Particle model and energy in matter

The Sun

Weight

Stars

Internal energy stored in materials

Gravity

Galaxies

Space physics

Seasons

Light years

MATTER AND SPACE PHYSICS
PRACTICE QUESTIONS

1. **a)** What is the weight of an object with a mass of 67kg on Earth? (1)

    Weight = mass × gravitational field strength (gravitational field strength = 10N/kg on Earth)

    b) The same object would weigh more on Saturn. Which of the following statements is the best explanation for this? (1)

    A. The gravitational field strength is lower on Saturn, due to its rings.

    B. The gravitational field strength is higher on Saturn, because it is in a different galaxy.

    C. The gravitational field strength on Saturn is lower, as it is further from the Sun.

    D. The gravitational field strength on Saturn is higher, because it has a larger mass than the Earth.

    c) Match the following features to their distance from the Earth. (2)

Uranus (a planet in our solar system)	7.8 light years
Wolf-359 (a star in our galaxy)	0.0003 light years
Large Magellanic Cloud (a galaxy outside the Milky Way)	163 000 light years

2. A sample of liquid water is heated until it boils and forms water vapour (a gas).

    a) Complete the blanks in the following explanation. (3)

    As the water is heated, the particles gain _____. This causes them to move _____. As the water boils, the particles move further away from each other. This means water vapour is less _____ than liquid water.

    b) Why does solid ice float on liquid water? Explain your answer in terms of particles. (2)

3. **a)** The average maximum temperature in December in England is 9.9°C, whilst the average maximum temperature in July is 20.9°C. Explain this difference in terms of the Earth's tilt. (2)

    b) On average it is much warmer in Australia in December than it is in the U.K. Why is this? (3)

    c) The Sun is a star. There are billions of other stars in our galaxy. Explain why the Earth's seasons are not affected by these other stars. (1)

NOTES

ANSWERS

Cells and organisation (pages 4–5)
Quick quiz
1. Light microscope
2. To carry out photosynthesis to produce glucose and oxygen.
3. Organs
4. False
5. Unicellular

The skeletal and muscular systems (pages 6–7)
Quick quiz
1. False
2. A pair of muscles which work in opposite directions to each other.
3. Two from: support, protection, movement, production of blood cells
4. A forcemeter
5. Lengthen

Nutrition and digestion (pages 8–9)
Quick quiz
1. Obesity
2. For the growth and repair of tissues.
3. Through their roots.
4. Enzymes
5. Small intestine

Gas exchange system in humans (pages 10–11)
Quick quiz
1. Through the stomata.
2. False
3. Trachea
4. Oxygen
5. The diaphragm relaxes and raises upwards.

Human reproduction (pages 12–13)
Quick quiz
1. Sperm
2. Oviduct
3. Uterus
4. Through the placenta.
5. The egg

Plant reproduction (pages 14–15)
Quick quiz
1. The transfer of pollen from the anther to the stigma of another flower.
2. It forms the seeds.
3. They help disperse the seeds.
4. Wind-pollinated flower.

Health (pages 16–17)
Quick quiz
1. The sharing of dirty needles.
2. It can cause people to see and hear things that are not really there.
3. Smoked or eaten.
4. Death from heart failure or suffocation. Damage to heart, liver and kidneys.

Practice questions: Structure and function of living organisms (pages 18–21)
1. **(a)** D
   (b) antagonistic (1); contracts (1); relaxes (1)
   (c) From: protection (i); support (i); producing blood cells (i)
2. **(a)** A insect pollination (i): B wind pollination (i)

(b)

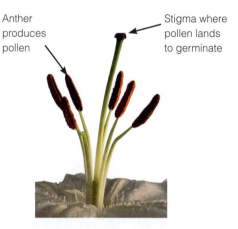

Anther produces pollen

Stigma where pollen lands to germinate

(c) Dispersal by animals (attached to their fur).
3. **(a)** No – plant cells and animal cells both contain a nucleus and mitochondria.
   (b) Two from: chloroplasts; cell wall; vacuole
   (c) An organism that has more than one cell.
4. **(a)** dietary fibre; proteins; carbohydrates
   (b) Obesity
   (c) Biological catalysts
5. **(a)** A
   (b) in; increases; decrease; increase
   (c) Smoking five cigarettes a day (1). Having asthma (1).

Photosynthesis (pages 22–23)
Quick quiz
1. Carbon dioxide
2. Oxygen
3. Chloroplasts
4. Chlorophyll

Cellular respiration (pages 24–25)
Quick quiz
1. Oxygen
2. Carbon dioxide
3. Microorganisms
4. Aerobic

Relationships in an ecosystem (pages 26–27)
Quick quiz
1. A plant.
2. Primary consumer/herbivore.
3. Secondary consumer/carnivore.
4. Heavy metals and pesticides.

Inheritance, chromosomes, DNA and genes (pages 28–29)
Quick quiz
1. Natural selection
2. species
3. Continuous
4. The process by which genetic information is passed from one generation to the next.
5. Watson, Crick, Franklin, Wilkins.

Practice questions: Biological processes (page 31)

1. **(a)**
   A. Grass
   B. Zebra/antelope
   C. Cheetah/lion
   (b) 2 marks from any of the following: Bees are pollinating insects (1). If there is a decline in the bee population, it could lead to a reduction in pollination of crops eaten by humans (1). This could lead to there being less food available (1).
   (c) That there are no living members of that species.
2. **(a)** Continuous variation
   (b) Natural selection
   (c) DNA
   (d) Gene banks could store embryos of the bird, allowing them to be reintroduced into the wild if they did become extinct.

The particulate nature of matter (pages 32–33)
Quick quiz
1. Solids
2. Gases
3. Solids
4. It will increase.
5. It will decrease.

Expansion and contraction (pages 34–35)
Quick quiz
1. Gain energy.
2. By cooling it.
3. They will lose energy/move slower.
4. When it is heated.
5. When it is cooled.

Atoms, elements and compounds (pages 36–37)
Quick quiz
1. 564g
2. Protons and neutrons
3. Negative
4. An element contains one type of atom. Compounds contain more than one type of atom.

Pure and impure substances (pages 38–39)
Quick quiz
1. From high concentration to low concentration.
2. The melting point will change (generally the melting point will decrease).
3. The solvent.
4. The solute.
5. The pore size of the filter.

Chemical reactions (pages 40–41)
Quick quiz
1. 7
2. Salt and hydrogen
3. Salt and water
4. Combustion
5. Oxidation

Energetics (pages 42–43)
Quick quiz
1. Exothermic reaction
2. Endothermic reaction
3. It will stay the same until the solid has completely melted.
4. It's releasing energy to the environment.
5. It will stay the same till the liquid has completely boiled.

The periodic table (pages 44–45)
Quick quiz
1. An alkali
2. Poor conductors
3. No, mercury is a liquid.
4. False
5. Groups

Materials (pages 46–47)
Quick quiz
1. Aluminium
2. No, iron is less reactive than carbon.
3. In iron ore or iron oxide.
4. A polymer is a large molecule made up of monomers joined together.
5. It is a composite.

Earth and atmosphere (pages 48–49)
Quick quiz
1. Mantle
2. They formed when magma cools slowly underground.
3. Metamorphic
4. Nitrogen
5. Climate change / greenhouse effect / global warming

Practice questions: Chemistry (pages 52–53)

1. **(a)**
   A. May contain fossils which aren't deformed – Sedimentary
   B. Formed from other rocks deep underground by high pressure and temperatures – Metamorphic
   C. Can be formed when magma cools on the earth's surface. – Igneous
   (b) Jasmine
   (c) Intrusive rocks form underground so the magma cools slowly (1), forming large crystals (1).

2. **(a)**

More reactive than A	Less reactive than A
Iron Aluminium	Copper

   (b) Carbon (1) could be used to extract the metal from its ore.
   (c) The unknown metal must be less reactive than carbon, because it is less reactive than iron, which is less reactive than carbon (1).
3. **(a)** Solute
   (b) Solvent
   (c) Filtration
   (d) A filter cannot be used (1), as the solid is dissolved in the liquid (1). Filters can only be used to separate mixtures where the solid is not dissolved in the liquid (1).
   (e) Evaporation

Calculation of fuel uses and costs in the domestic context (pages 54–55)
Quick quiz
1. Energy transferred over a period of time.
2. kWh
3. From: coal, oil or gas.
4. Become overweight/suffer from obesity.
5. From: It does not provide a constant supply of electricity; many people believe wind turbines are ugly and spoil natural habitats.

Energy changes and transfers (pages 56–57)
Quick quiz
1. Radiation
2. Conduction
3. From gravitational potential energy to kinetic energy.
4. To reduce the amount of heat lost from the house.
5. From chemical to electrical.

Changes in systems (pages 58–59)
Quick quiz
1. Chemical
2. Light and heat
3. By climbing above the surface of the Earth, for example using stairs or a ladder.
4. Electrical energy flows into it.

Practice questions: Energy (page 61)

1. **(a)** January – March
   (b) £310.24
   (c) It would reduce the bill.
   (d) Insulation would reduce heat lost from the house (1) and

therefore less energy would be required to heat the house and maintain the desired temperature (1).

2. **(a)** Climate change / greenhouse effect / global warming (accept acid rain)
    (b) They are used faster than the millions of years it takes to form them.
    (c) Any two from: wind power, solar power, wave power, tidal power, geothermal power, biomass.
    (d) The wind does not blow all the time (1), so electricity will not always be generated.

Describing motion (pages 62–63)
Quick quiz
1. Speed = distance / time
2. 10.18 m/s
3. 62.5 seconds
4. 90m

Forces, motion and balanced forces (pages 64–65)
Quick quiz
1. It will begin to move in the direction of the largest force.
2. It will continue to move in the same direction at the same speed.
3. Air resistance
4. From: gravity, magnetism and static electricity.
5. Newtons

Pressure in fluids (pages 66–67)
Quick quiz
1. The pressure on the object increases.
2. False
3. At right angles to the surface (along the normal).
4. Upthrust
5. Pressure is the ratio of force over the area it is applied.

Practice questions: Motion and forces (page 69)
1. **(a)** All the forces on the ship are balanced (1).
    (b) Upthrust balances the weight of the ship, which allows it to float (1).
    (c) The submarine will be diving beneath the surface of the water (1). As it gets deeper, the pressure increases due to the weight of the water above the submarine (1). High pressure could damage the submarine (1).
2. **(a)** 17m/s
    (b) 11.1 seconds
    (c) (i) 0–10 minutes (ii) 20–30 minutes

Observed waves, sound waves and energy and waves (pages 70–71)
Quick quiz
1. Frequency
2. Longitudinal
3. Dogs
4. Smooth flat surfaces
5. Solid

Light waves (pages 72–73)
Quick Quiz
1. Yes
2. It refracts.
3. It will split into different colours.
4. Specular reflection
5. All colours except blue.

Practice questions: Waves (page 75)
1. **(a)** The dog whistle produces a high frequency sound (i). Dogs have a greater auditory range than humans, so they can hear high pitched sounds that humans cannot (i).
    (b) Transverse waves
    (c) C
2. **(a)** Light travels faster than sound.
    (b) B
    (c) Light can travel through a vacuum (1) / sound cannot travel through a vacuum (1).

Current electricity (pages 76–77)
Quick quiz
1. Ohms

2. Volts
3. High resistance
4. False
5. The flow of charge through a circuit.

Static electricity and magnetism (pages 78–79)
Quick quiz
1. A D.C. motor is made.
2. They attract each other.
3. Increasing the number of turns on the coil or increasing the current.
4. The magnetic field will not be produced.
5. The same shape as the magnetic field around a bar magnet.

Practice questions: Electricity and electromagnetism (page 81)
1. **(a)** D
    (b) positively; negatively
    (c) Wrap the wire around an iron core.
    (d) A compass
    (e) The same shape as a bar magnet.
2. **(a)** A = Series circuit; B = Parallel circuit
    (b) a. potential difference   ii. Volts
         b. Current   iii. Amperes
         c. Resistance   i. Ohms

Physical changes (pages 82–83)
Quick quiz
1. A change of state where a solid becomes a gas, without first forming a liquid.
2. When a solute dissolves in a solvent to form a solution.
3. Gases and liquids
4. Decrease the volume of the container it is in.
5. From a high concentration to a low concentration.

Particle model and energy in matter (pages 84–85)
Quick quiz
1. A gas
2. In a liquid
3. In a liquid state
4. Chemical

Space physics (pages 86–87)
Quick quiz
1. The distance light travels in one year.
2. Winter
3. Planet, solar system, galaxy, universe.
4. Summer
5. The Sun has a greater mass than the Earth.

Practice questions: Matter and space physics (page 89)
1. **(a)** 670N
    (b) D
    (c) Uranus– 0.0003 light years
         Wolf-359 – 7.8 light years
         Large Magellanic Cloud – 163 000 light years
2. **(a)** energy; faster; dense
    (b) Ice is less dense than liquid water (1)/because the particles are further apart (1).
3. **(a)** In December, England is tilted away from the Sun (1), so it is less directly exposed to the Sun (1).
    (b) Australia is in the southern hemisphere, so is tilted towards the Sun in December (1). It is therefore more directly exposed to energy from the Sun (1).
    (c) They are too far away.

PERIODIC TABLE

Key

Metals

Non-metals

Atomic Number → 1
Atomic Symbol → **H**
Name → hydrogen

1	2

3 **Li** lithium	4 **Be** beryllium
11 **Na** sodium	12 **Mg** magnesium

19 **K** potassium	20 **Ca** calcium	21 **Sc** scandium	22 **Ti** titanium	23 **V** vanadium	24 **Cr** chromium	25 **Mn** manganese	26 **Fe** iron	27 **Co** cobal
37 **Rb** rubidium	38 **Sr** strontium	39 **Y** yttrium	40 **Zr** zirconium	41 **Nb** niobium	42 **Mo** molybdenum	43 **Tc** technetium	44 **Ru** ruthenium	45 **R** rhodiu
55 **Cs** caesium	56 **Ba** barium	57 **La** lanthanum	72 **Hf** hafnium	73 **Ta** tantalum	74 **W** tungsten	75 **Re** rhenium	76 **Os** osmium	77 **Ir** iridiu
87 **Fr** francium	88 **Ra** radium	89 **Ac** actinium	104 **Rf** rutherfordium	105 **Db** dubnium	106 **Sg** seaborgium	107 **Bh** bohrium	108 **Hs** hassium	109 **Mt** meitner

0 or 8

3	4	5	6	7	2 **He** helium
5 **B** boron	6 **C** carbon	7 **N** nitrogen	8 **O** oxygen	9 **F** fluorine	10 **Ne** neon
13 **Al** aluminium	14 **Si** silicon	15 **P** phosphorus	16 **S** sulfur	17 **Cl** chlorine	18 **Ar** argon

29 **Ni** nickel	29 **Cu** copper	30 **Zn** zinc	31 **Ga** gallium	32 **Ge** germanium	33 **As** arsenic	34 **Se** selenium	35 **Br** bromine	36 **Kr** krypton
Pd adium	47 **Ag** silver	48 **Cd** cadmium	49 **In** indium	50 **Sn** tin	51 **Sb** antimony	52 **Te** tellurium	53 **I** iodine	54 **Xe** xenon
Pt atinum	79 **Au** gold	80 **Hg** mercury	81 **Tl** thallium	82 **Pb** lead	83 **Bi** bismuth	84 **Po** polonium	85 **At** astatine	86 **Rn** radon
Ds nstadtium	111 **Rg** roentgenium							

INDEX